JOHN D. LENK
Consulting Technical Writer

HANDBOOK OF OSCILLOSCOPES
Theory and Application
REVISED AND ENLARGED

PRENTICE-HALL, INC., Englewood Cliffs, New Jersey 07632

Library of Congress Cataloging in Publication Data

Lenk, John D.
    Handbook of oscilloscopes.

    Includes index.
    1. Cathode ray oscilloscope. I. Title.
TK7878.7.L38 1982   621.37'47   81–10680
ISBN 0–13–380576–X   AACR2

Editorial/production supervision and interior design
    by Barbara A. Cassel
Manufacturing buyer: Gordon Osbourne

Printed in the United States of America

10 9 8 7 6 5 4 3 2 1

ISBN 0-13-380576-X

Prentice-Hall International, Inc., *London*
Prentice-Hall of Australia Pty. Limited, *Sydney*
Prentice-Hall of Canada, Ltd., *Toronto*
Prentice-Hall of India Private Limited, *New Delhi*
Prentice-Hall of Japan, Inc., *Tokyo*
Prentice-Hall of Southeast Asia Pte. Ltd., *Singapore*
Whitehall Books Limited, *Wellington, New Zealand*

To *Irene,* the most Wonderful Wife in the World;
and to *Lambie,* our little Wookie

Contents

Chapter 6
MEASURING VOLTAGE AND CURRENT 78

Chapter 7
MEASURING TIME, FREQUENCY, AND PHASE 111

Chapter 8
USING OSCILLOSCOPES WITH SWEEP GENERATORS 138

Preface

This revised edition of the *Handbook of Oscilloscopes: Theory and Application* carries through all the features that made the first edition such an outstanding success. That is, the revised edition bridges the gap between oscilloscope theory and practical applications. The new edition thus serves the dual purpose of serving as a basic textbook for student technicians, hobbyists, and experimenters, and as a factual guidebook for experienced working technicians and engineers. All the chapters in the revised edition have been expanded to include new material. Existing information has been updated to reflect present-day trends, especially in the extensive use of curve tracers. Also, much of the material from the first edition has been revised for clarification and/or simplification.

Manufacturers of oscilloscopes provide instruction manuals on the operation of and circuit theory for their particular instruments. Rarely, however, do these manuals describe all of the many applications for which the modern oscilloscope can be used. Because the oscilloscope has been in such common use for many years, it is generally assumed that the operator will know "automatically" the procedures for using oscilloscopes. This is generally not the case. For example, how do you test transistors in-circuit? Or how do you measure the slew rate of an amplifier, or monitor the modulation pattern of an SSB transmitter?

As in the original, this edition of the handbook fills in these information gaps and can be used to supplement the operating instructions of any oscilloscope, whether it be a low-cost shop type or a precision laboratory instrument. This is done by providing a variety of test, measurement, service, and troubleshooting procedures using the oscilloscope as the basic tool. These procedures are presented in "cookbook" fashion. Each procedure is preceded by a brief description of the "why" and "where" for the particular test. These descriptions offer a digest to readers who may be unfamiliar with some specialized oscilloscope applications, and want to put the step-by-step procedures to immediate use. Each operation is illustrated with test connection diagrams. Although every possible use of an oscilloscope has not been included, all the practical, experience-proven applications are here.

Assuming that some readers are not familiar with the operating principles and characteristics of oscilloscopes, the initial chapters give *simplifed* presentations of these details. Chapters 1 through 4 cover oscilloscope basics —typical operating controls, specifications, and performance—as well as a brief description of oscilloscope accessories. Throughout, the descriptions are kept to the block diagram or simplified level. Unnecessary and elaborate circuit descriptions are avoided.

Many professionals have contributed their talent and knowledge to the revision and enlargement of the new edition. The author gratefully acknowledges that the tremendous effort to make this new edition such a comprehensive work is impossible for one person, and he wishes to thank all who have contributed directly and indirectly. The author wishes to give special thanks to the following: B&K Precision Test Instrument Product Group of Dynascan Corporation, Heathkit, Hewlett-Packard, General Electric, Motorola, Tektronix, Texas Instruments, RCA Corporation Solid-State Division, and the Society of Motion Picture and Television Engineers. The author also wishes to thank Mr. Joseph A. Labok of Los Angeles Valley College for his help and encouragement.

John D. Lenk

1

Oscilloscope Basics

The cathode-ray oscilloscope (CRO) is an extremely fast X-Y plotter capable of plotting an input signal versus another signal, or versus time, whichever is required. A luminous spot acts as a "stylus" and moves over the display area in response to input voltages. In most applications, the Y-axis (vertical) input receives its signal from the voltage being examined, moving the spot up or down in accordance with the instantaneous value of the voltage. The X-axis (horizontal) input is usually an internally generated linear ramp voltage which moves the spot uniformly from left to right across the display screen. The spot then traces a curve which shows how the input voltage varies as a function of time.

If the signal under examination is repetitive, at a fast enough rate, the display appears to stand still. The CRO is thus a means of visualizing time-varying voltages. As such, the oscilloscope has become a universal tool in all kinds of electronic investigations. Oscilloscopes operate on voltages. It is possible, however, to convert current, strain, acceleration, pressure, and other physical quantities into voltages by means of transducers, and thus to present visual representations of a wide variety of dynamic phenomena on oscilloscopes.

The formal name *cathode-ray oscilloscope* is usually abbreviated to *oscilloscope* or, simply, *scope*. An *oscillograph* is the pictoral representation of an oscilloscope trace. Some older texts apply the word "oscillograph" to the complete equipment.

1-1.   THE CATHODE-RAY TUBE

All circuits of an oscilloscope are built around a cathode-ray tube (CRT). Figure 1-1 shows the internal construction of a typical tube of this type. As with other vacuum tubes, the filament heats the cathode to the degree where it emits electrons. The control grid influences the amount of current flow, as in standard vacuum tubes. Two anodes are used, each having a positive d-c potential applied to it. These anodes accelerate the electrons and form them into a beam. The intensity of the beam is regulated by the potential applied to the control grid.

The cathode consists of a nickel cylinder with an emitting element fused at its end. This element is made of either barium or strontium oxide and permits a release of sufficient electrons for an electron stream to form.

The grid structure, although controlling electron flow as in conventional tubes, differs from the wire mesh of receiving tubes and consists of a cylinder with a tiny circular opening, to keep the electron stream small enough. The beam is focused into a sharp pinpoint by controlling the voltage on the first anode. The two anodes of the cathode-ray tube can be compared to a glass

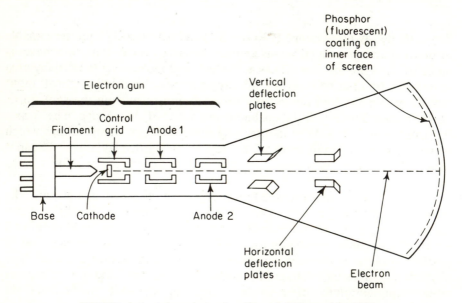

FIGURE 1-1   Internal construction of CRT used in CROs.

lens system, such as used in movie projectors, because the anodes focus the beam to a pinpoint at the face of the tube. A high voltage is applied to the second anode so that the electron stream will attain high velocity for increased intensity and visibility when it strikes the tube face. The beam-forming section of the tube is known as an *electron gun.*

The inside of the tube face is coated with phosphor so that, when electrons strike this coating, it will *fluoresce* and emit light. After the exciting electron stream has left the area, the fluorescing characteristics which emit light rapidly decay, and the light level is reduced. The chemical composition of the coating, however, can be such that the emitted light persists for an appreciable interval, so that one can observe the light visually. Since the beam is swept across the tube at a fairly rapid rate during normal operation, the light must persist for a time interval sufficient for it to leave a complete trace of the waveform drawn on the tube face. At the same time, the persistence of the phosphorescent coating should be sufficiently short so that, if the beam stops, the pattern traced on the tube will disappear very rapidly.

A standard numbering system is used, within the tube designation code, for ready identification of the phosphor characteristics. For example, if 3AP1 is a cathode-ray tube's numerical designation, the tube has a 3-in. face, because the first number identifies the face diameter. The P1 designation indicates a medium-persistence phosphor, which has a green glow when excited. The A designation refers to the internal construction and indicates that this particular tube has some structural changes with respect to a 3P1 cathode-ray tube.

1-2. BEAM DEFLECTION SYSTEM

As shown in Fig. 1-1, two sets of plates are present within the tube, beyond the second anode. These plates are for deflecting the electron beam both horizontally and vertically so that the beam will "write out" information delivered to the deflecting system. Such a system is known as *electrostatic deflection* and is predominantly used in oscilloscopes. Magnetic deflection, once used in a few oscilloscopes, is more often used in TV picture tubes and radar displays.

The electrostatic beam deflection is accomplished through the two pairs of parallel plates. For example, a voltage applied across the horizontal deflection plates will influence the beam, because the negative potential on one of the horizontal plates repels the electron stream, whereas the positive potential on the other horizontal deflection plate attracts the beam. If such a voltage is a sawtooth type, the gradually rising potential of the sawtooth will gradually pull the beam toward the positive horizontal deflection plate. Therefore, the electron beam is made to scan across the face of the tube. Also, any potential applied to the vertical deflection plates will cause the beam to move vertically.

Figure 1–2 shows how an oscilloscope traces out a sine wave on the tube face, when such a signal is applied to the vertical input of the oscilloscope. In Fig. 1–2a, the sine wave applied at the vertical input of the oscilloscope is shown. If the internal horizontal sweep generator is turned off, the rising positive potential of the first alternation of the input signal causes the electron beam to move upward, as shown in Fig. 1–2b. When the negative alternation of the input signal arrives at the vertical deflection plates, the electron beam is pulled downward from the center, also as shown in Fig. 1–2. The rapid rise and fall of the signal alternation causes the electron beam to move up and down the face of the cathode-ray tube very rapidly, leaving a vertical line trace.

If the horizontal oscillator within the oscilloscope is now turned on, a sawtooth of voltage, as shown in Fig. 1–2c, will be applied to the horizontal deflection plates. The rising potential of the sawtooth voltage causes the right horizontal plate to become positive and the left horizontal plate to become negative. The negative left plate and the positive right plate cause the beam to move from left to right, because the beam is repelled by the negative plate and attracted by the positive plate. If one sawtooth occurs for each cycle of the sine-wave signal, the beam will be pulled across the face of the tube once for each cycle. Thus, as the first alternation of the input cycle rises in amplitude, the electron beam would normally rise in a vertical plane, as shown in Fig. 1–2b. The sawtooth on the horizontal plates gradually pulls the beam from left to right and traces out the input signal waveshape in visual form.

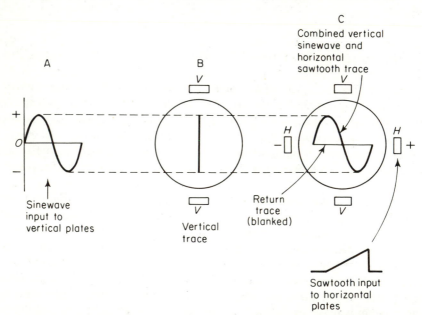

FIGURE 1-2 Vertical sine wave and horizontal sawtooth sweep produce CRO trace.

1-3. BASIC FREQUENCY MEASUREMENT

Waveshapes of square waves, pulses, or any other types of signals can be observed on the face of the oscilloscope screen. If the input signal has a frequency twice that of the sawtooth applied to the horizontal plates, two cycles will appear on the screen, because the beam is pulled across the screen only once for each two cycles of the input signal. By regulating the ratio of the input signal frequency to the sawtooth sweep frequency, portions of the input signal, or a number of cycles of the input signal, can be made visible on the screen. By calibrating the frequency of the horizontal sweep waveform so that its exact frequency is known, the frequency of the input signals to the oscilloscope can be calculated. For example, if four cycles of a sine wave appear on the screen and the sawtooth generator is set for 100 cycles, the frequency of the signal applied at the input of the oscilloscope is 400 cycles.

1-4. BASIC VOLTAGE MEASUREMENT

In addition to frequency measurements, the oscilloscope can be used for reading the peak-to-peak voltages of a-c signals, pulses, square waves, and so on, as well as d-c voltages. For reading voltages, a transparent plastic screen is attached to the face of the oscilloscope. Such a screen (also known as *grid, grid mask, mask, grating, graticule*) is found in several forms. Usually, a screen is a transparent scale with vertical and horizontal lines spaced one division apart. The screen is fitted against the screen of the CRT. This allows time and amplitude to be read directly. These graduated scales often have small markings which subdivide the major divisions to assist in making accurate measurements. Most laboratory oscilloscopes are set up in centimeters. Some shop oscilloscopes are calibrated in inches. Still other oscilloscopes use no particular standard of measurement, but are simply equal-spaced "divisions." *In all the procedures throughout this book it is assumed that the oscilloscope is calibrated in centimeters.*

To calibrate the oscilloscope screen, an a-c voltmeter of known accuracy is used initially. A low-voltage a-c signal must be present for calibration. Some oscilloscopes have a terminal on the front panel which supplies such an a-c reference voltage. For example, if the reference voltage is 5 V, this voltage is applied to the vertical input terminal of the oscilloscope. The internal horizontal sweep generator is shut off so that a vertical trace, as shown in Fig. 1-2b, is visible on the screen. This vertical line represents the peak-to-peak voltage of the input signal. The vertical height control is then adjusted so that the line is five divisions high, and the control is left in this position after calibration. Knowing the rms value of the applied a-c calibrating voltage, the peak-to-peak voltage can be calculated by multiplying the rms value by 1.41 to obtain the peak value, then doubling the peak value to obtain the peak-to-peak value.

1-5. BASIC OSCILLOSCOPE CIRCUITS

Figure 1-3 is a block diagram of a *typical* oscilloscope. It would be almost impossible, and beyond the scope of this book, to describe all the circuits in modern oscilloscopes. Many of these circuits are special purpose. The basic circuits are used in various combinations. Instead of attempting a description of every circuit and combination, the following discussion covers a *complete* working oscilloscope.

1-5-1. Vertical (Y-Axis) Channel

Signals to be examined are usually applied to the vertical, or *Y*, deflection plates through the vertical amplifier. A vertical deflection amplifier is required, since the signals are usually not strong enough to produce measurable vertical deflection on the CRT. A typical CRT requires 50 V dc to produce a deflection of 1 in. The high-gain amplifier in a laboratory type of oscilloscope permits a 1-in. deflection with 0.5 mV; the average shop type of oscilloscope requires at least 20 mV for a 1-in. deflection. Another difference between laboratory and shop amplifiers is the frequency response. As in the case of any amplifier, the frequency response must be wide enough to pass faithfully the *entire band* of frequencies to be measured on the oscilloscope. For simple audio work, an upper limit of 200 kHz is sufficient. Television service requires a passband of at least 5 MHz. Some laboratory oscilloscopes provide up to 1000 MHz (1GHz); 100 MHz is common for laboratory scopes.

When high-voltage signals are to be examined, they can be applied directly to the vertical deflection plates on most oscilloscopes.

The vertical amplifier output is also applied to the sync amplifier through the sync selector switch in the internal position. This permits the horizontal sweep circuit to be triggered by the signal being examined.

1-5-2. Horizontal (X-Axis) Channel

Usually, the horizontal deflection plates are fed a sweep voltage that provides a time base. As in the case of the vertical channel, the horizontal plates are fed through an amplifier, but they can be fed directly when the voltages are of sufficient amplitude. When external signals are to be applied to the horizontal channel, they can also be fed through the horizontal amplifier, via the sweep selector switch in the external position. When the sweep selector is in the internal position, the horizontal amplifier receives an input from the sawtooth sweep generator which is triggered by the sync amplifier.

In most oscilloscopes, the sawtooth sweep voltage is generated by a multivibrator, relaxation oscillator, or pulse generator. There are four basic types of sweeps. The *recurrent* sweep presents the display repetitively, and the eye sees a lasting pattern. In a *single sweep,* the spot is swept once across the screen in response to a trigger signal. The trigger can be obtained from the

6

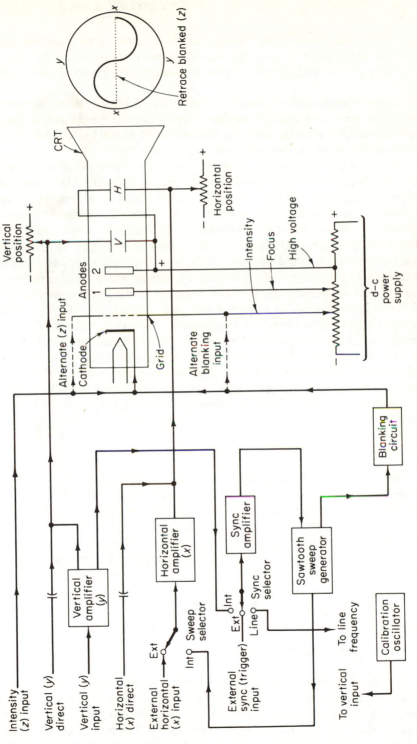

FIGURE 1-3 Simplified block diagram of typical oscilloscope.

7

signal under study, or from an external source. In most cases, a *driven sweep* is used where the sweep is recurrent but triggered by the signal under test. In special cases, some oscilloscopes provide a *nonsawtooth sweep,* such as a sine wave.

Sweep frequencies vary with the type of oscilloscope. A laboratory oscilloscope may have sweep frequencies up to several megahertz; a simple shop oscilloscope for audio work has an upper limit of 100 kHz. Most TV service requires a horizontal sweep frequency up to 1 MHz.

Whatever type of sweep is used, it must be synchronized with the signal being investigated. If not, the pattern will appear to drift across the screen in a random fashion. Hence, a sync system is needed. Three usual sources for synchronization can be selected by the sync selector: internal, where the trigger is obtained from the signal under investigation (through the vertical amplifier); external, where an external trigger source is also used to trigger or initiate the signal being measured; and line, where the sync trigger is obtained from the line frequency. Line sync is often used in TV service where an external sweep generator and an oscilloscope are both triggered at the line frequency.

The oscilloscope sweep system is also used to produce a *blanking* signal. The blanking signal is necessary to eliminate the retrace that would occur when the sweep trace snaps back from its final (right-hand) position to the initial or starting point. This retrace could cause confusion if it were not eliminated by blanking the CRT during the retrace period with a high negative voltage on the control grid (or a high positive voltage on the CRT cathode). The blanking voltage is usually developed (or triggered) by the sweep generator.

1-5-3. Intensity (Z-Axis) Channel

Intensity modulation, sometimes known as *Z-axis modulation,* is accomplished by inserting a signal between ground and the cathode (or control grid) of the CRT. When the signal voltage is large enough it can cut off the CRT on selected parts of the trace, just as the retrace blanking signal does. Z-axis modulation is applied during the normally visible portion of the trace.

The Z-axis can also be used to brighten the trace. Periodically applying positive pulse voltages to the CRT control grid (or negative pulses to the cathode) brightens the electron beam throughout its trace to give a third, or Z, dimension. These periodically brightened spots can be used as markers for time calibration of the main waveform.

1-5-4. Positioning Controls

For many measurements, it is necessary to provide some means of positioning the trace on the CRT face. Such centering provisions are accomplished by applying small, independent, internal d-c potentials to the deflection plates and controlling them by means of potentiometers. The centering or positioning controls are particularly useful during voltage calibration, and during enlargement of a waveform for examination of small characteristics. The portion of interest may move off the CRT screen so that positioning controls are necessary to bring it back on again.

1-5-5. Focus and Intensity Controls

The CRT electron beam is focused in a manner similar to that of an optical lens, but the focal length is altered by changing the ratio of potentials between the first and second anodes. This ratio is changed by varying the potential on the anode with the focus control, which is a potentiometer usually located on the oscilloscope front panel. The potential on the second anode remains constant.

The intensity of the beam is varied by the intensity control potentiometer which changes the grid potential with respect to the cathode, thus permitting more or fewer electrons to flow. Because the potentials applied to the control grid and to both anodes are taken from a common voltage divider network, any change made in the setting of the intensity control requires a compensating change in the setting of the focus control, and vice versa.

1-5-6. Calibration Circuit

Most laboratory oscilloscopes have an internally generated and stabilized waveform of known amplitude which serves as a calibrating reference. Usually, a square wave is used, with the calibrating signal accessible on a front-panel connector.

On shop oscilloscopes, the 60-Hz line voltage is used. Either way, the calibrating voltage can be applied to the vertical channel by running a lead between the front-panel calibration output connector and the vertical input connector. On some oscilloscopes, the calibrating voltage is applied to the vertical input through a front-panel control switch.

1-5-7. Other Oscilloscope Circuits

Those familiar with oscilloscope circuits will note that the discussion has not covered many basic controls, such as amplifier gain or attenuation, astigmatism, sweep frequency, time, sync polarity, trigger level, and slope, to name a few. These controls, as well as all other controls of typical oscilloscopes, are covered from the operator's viewpoint in Chapter 2.

1-6. STORAGE OSCILLOSCOPE

The storage oscilloscope is especially useful for one-shot displays, where it is not practical to photograph the display. The storage oscilloscope principle is also used in *computer graphics*. A storage oscilloscope holds a display on the screen for an indefinite time. Several waveforms may be superimposed for comparison. When desired, the operator can remove the display by depressing an erase button. A storage oscilloscope can be used as a conventional oscilloscope when the storage function is disabled.

Several types of storage cathode-ray tubes are available. Storage CRTs may be classified as either *bistable* or *halftone* tubes. The stored display on a bistable tube has one level of brightness. A halftone tube has the capacity of displaying a stored signal at different levels of brightness. The brightness of a halftone tube depends on beam current and the time the beam remains on a particular storage element. A bistable tube either stores or does not store an event. All stored events have the same brightness.

Storage CRTs may also be classified as either *direct-viewing* or *electrical-readout* tubes. An electrical-readout tube has an electrical input and output. A direct-viewing tube has an electrical input but a visual output.

Figure 1–4 is a simplified diagram of a typical storage CRT. The focus, intensity, and accelerator electrodes have been omitted for clarity.

In addition to the electron gun (known as the *writing gun* of a conventional CRT), the storage CRT also has a *flood gun* (or guns). Unlike the conventional CRT, a storage CRT has four screens behind the display area.

The *viewing screen* is a phosphor screen similar to that of a conventional CRT. Directly behind the viewing screen is a *storage screen,* usually in the form of a fine metal mesh coated with dielectric. A *collector screen* and *ion-repeller screen* are located behind the storage screen. Not all storage CRTs have the ion-repeller screen.

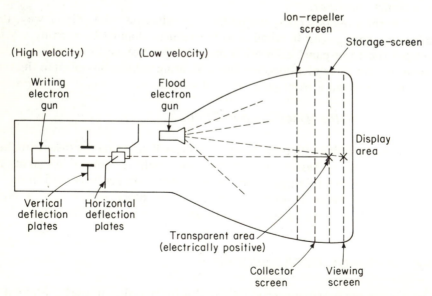

FIGURE 1-4 Simplified diagram of storage cathode-ray tube.

When signals are applied to the horizontal and vertical deflection plates, the electron beam from the writing gun is moved to trace out the corresponding waveform, in the normal manner. When this beam strikes the storage screen, an electrically positive trace is written out, by removing electrons from the storage screen dielectric atoms where it strikes. This leaves a positive trace stored on the storage screen after the signal has been removed. The excess electrons removed from the storage screen are collected by the collector screen.

The removal of electrons from the storage screen makes the affected areas (of the dielectric) transparent to low-velocity electrons (that is, low-velocity electrons pass through the areas written out by the high-velocity electron beam). Any time after the trace has been written on the storage screen, the flood gun can be turned on to spray the entire storage screen with low-velocity electrons. These electrons pass through the transparent areas but do not penetrate other areas of the storage screen. Usually, the storage screen is maintained at the same voltage level as the cathode of the flood gun, thus repelling the flood gun electrons (except for the transparent areas).

Electrons passing through the storage screen reproduce the trace on the phosphor viewing screen. In effect, the storage screen acts as a "stencil" to the low-velocity electrons from the flood gun. When it is desired to erase the display, the voltage at the collector screen is changed, causing the storage screen to discharge (all of the electrons are restored to the dielectric atoms in the storage screen).

1-7. SAMPLING OSCILLOSCOPE

Conventional oscilloscopes are limited in bandwidth to frequencies in the megahertz region. Sampling oscilloscopes have bandwidths up to about 12 GHz (and above). This permits the sampling oscilloscope to measure signals of extremely high frequency or to monitor waveforms of very fast rise times (0.5 ns or faster is typical).

The sampling oscilloscope uses a stroboscopic approach to reconstruct the input waveform from samples taken during many recurrences of the waveform. This technique is illustrated in Fig. 1-5. In reconstructing a waveform, the sampling pulse "turns on" the sampling circuit for an extremely short interval, and the waveform voltage at that instant is measured. The CRT spot is positioned vertically to correspond to this voltage amplitude.

The next sample is taken during a subsequent cycle at a slightly later point on the input waveform. The CRT spot moves horizontally a short distance and is repositioned vertically to the new voltage. In this way, the oscilloscope plots the waveform point by point (or dot by dot). As many as 1000 samples are used to reconstruct a single, recurrent waveform.

A bright trace is obtained regardless of sampling rate, sweep speed, or

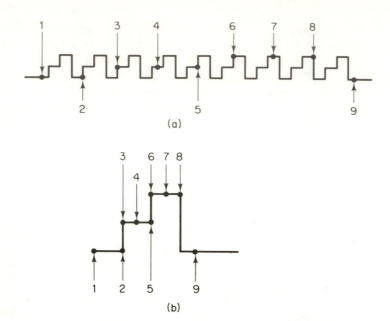

FIGURE 1-5 Sampling oscilloscope/stroboscopic technique to reconstruct a waveform from many samples. (a) Samples taken during nine recurrences of the same waveform; (b) reconstructed waveform.

waveform duty cycle, since each CRT spot remains "on" during the full interval between samples. For example, assume that a pulse with an extremely fast rise time is monitored. If the rise time is at the maximum speed of the electron beam in a conventional oscilloscope, the beam moves so quickly that it produces a weak trace on the leading and trailing edges of the waveform. In the sampling oscilloscope, the leading and trailing edges are sampled many times at different points from top to bottom, thus producing a bright, steady trace.

The *random sampling oscilloscope* is an improved version of the sampling oscilloscope. Operation of a random sampling oscilloscope can be divided into two separate parts: (1) timing the sample to fall somewhere within a time window (displayed portion of the waveform or signal period), and (2) constructing the display from a series of such samples placed at random. In a non-random-sampling oscilloscope, the samples are taken at a specific time interval. In random sampling, there is considerable time uncertainty between the signal being sampled and the sample-taking process.

Figure 1–6 shows how the X and Y deflection signals are generated in a random sampling oscilloscope. As shown, five randomly placed samples are taken from a signal. Note that these samples are taken of *successive repetitions* of the signal. This is known as random *equivalent-time* sampling. There is also a method of random *real-time* sampling, where many randomly placed samples are taken during a single, relatively slow, signal occurrence.

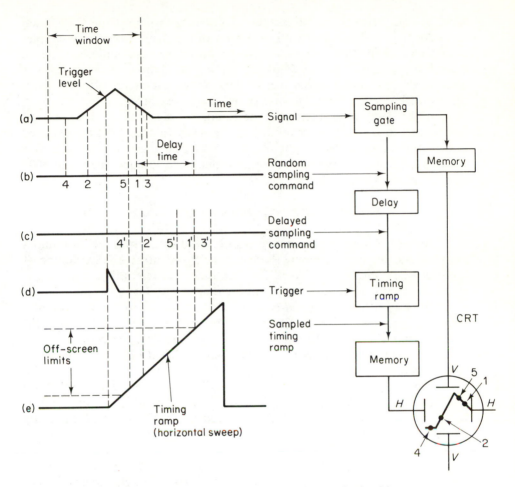

FIGURE 1-6 How X and Y deflection signals are obtained in random sampling oscilloscopes.

The Y component of the first sample is passed through a sampling gate and held in a memory circuit for a brief period of time. The Y signal is then applied to the vertical deflection plates in the normal manner, deflecting the CRT spot vertically.

The sampling command (which opened the sampling gate) is delayed by a fixed interval, as indicated in Fig. 1-6c. The delayed sampling command is used to sample a horizontal timing ramp (sawtooth sweep). The timing ramp is triggered when the input signal reaches a certain level, as shown in Fig. 1-6a and d. The sample of the timing ramp is held in a memory circuit for a brief period and then applied to the horizontal deflection plates in the normal manner, deflecting the CRT spot horizontally.

The process is repeated for each of the remaining samples. Each sample

13

supplies both vertical and horizontal information to deflect the CRT beam from dot to dot, thus constructing a display of the signal from those samples that fall within the "time window." Note that sample 3 is not displayed, even though it is sampled and available at the vertical deflection plates. This is because the horizontal sweep has already driven the beam off the screen (horizontally) before the horizontal sweep ramp is sampled.

If the delay time interval is increased, a time shift of the sampling distribution to the left (earier in time) is necessary in order that the required information be collected for the display. Also note that although only five samples are shown, a typical random sampling oscilloscope might use 1000 samples to reconstruct such a waveform.

1-7-1.   Random Sampling Oscilloscope Operation

Figure 1–7 is a complete operational block diagram of a typical random sampling oscilloscope, including those circuits that control the distribution of samples across the time window. The following is a description of the basic circuit functions, referenced to the timing diagram of Fig. 1–6.

The *trigger recognizer and hold-off* block responds to the presence of a suitable trigger (when the input signal has reached a certain level, Fig. 1–6a) and immediately starts the timing ramp. The hold-off function prevents restarting the timing ramp until the ramp has sufficient time to complete the previous cycle.

The *timing ramp* is a linear ramp with the slope controlled by a *time/division switch*. The starting and subsequent sampling of this ramp is shown in Fig. 1–6e. The sampled and stored levels of the timing ramp are available at the output of the *horizontal memory,* where they supply the horizontal (*X*-axis) data to the display.

The *ratemeter* block also receives an input indicating that a trigger has been recognized (the input signal has reached the trigger level) and proceeds to measure the repetition rate of such recognitions over an extremely wide range of trigger repetition rates. On the basis of this measurement and an error signal supplied by the *differential comparator,* the ratemeter then supplies the *slewing ramp* with a pretrigger signal. This pretrigger is the ratemeter's "best guess" as to when to start the slewing ramp. The pretrigger may contain considerable time uncertainty on a sample-to-sample basis. If a number of successive samples are taken later than desired, an appropriate error signal arrives to cause the ratemeter to start the slewing ramp earlier for the next few cycles.

The *slewing ramp* is a linear ramp that is started on command from the ratemeter. The slope of the slewing ramp is controlled by the time/division switch. The *slewing comparator* provides both the sampling commands and the delayed sampling commands, shown in Fig. 1–6b and c. The delayed sampling command is issued when the relatively fast slewing ramp reaches the

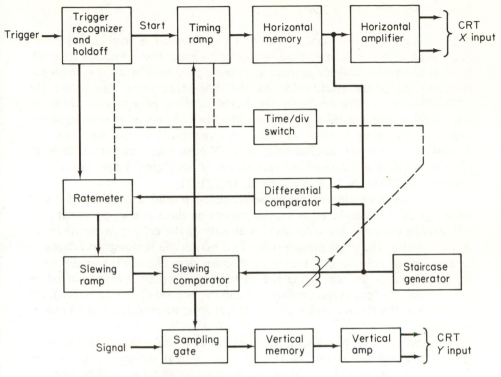

FIGURE 1-7 Operating block diagram of a typical random sampling oscilloscope.

voltage level of the *staircase generator* output. Successive sweeps of the slewing ramp find the staircase at slightly higher levels. Thus, the resulting comparisons and sampling commands are successively delayed, or "slewed" in time.

The delayed sampling commands are generated in a similar fashion from the slewing ramp and staircase, but a *d-c offset* added to the staircase causes these comparisons to occur later by the fixed time interval (Fig. 1-6c).

The *differential comparator* receives both the horizontal signal and the staircase, and generates an error signal when the horizontal signal does not track along with the staircase on the basis of an *average* of many samples. The staircase generates its signal at a constant rate. The horizontal signal depends on the average of the samples taken. Thus, a "closed-loop" system is formed, causing a random sampling distribution to slew across the time window under control of the staircase generator. The output of the slewing comparator sends a command to the sampling gate, permitting the signal to pass to the vertical memory circuit, as shown in both Figs. 1-6 and 1-7.

1-8. THE VECTORSCOPE

A vectorscope is used in service of color TV receivers. The vectorscope is normally used in conjunction with a *keyed rainbow generator* to check a color TV receiver's response to color signals. The colors produced by a TV receiver depend on the phase relationship of the 3.58-MHz *color signal* and the 3.58-MHz *color reference burst*. The keyed rainbow generator produces 10 color signals, each spaced 30° apart, resulting in a display of 10 corresponding colors (arranged as vertical stripes or a "rainbow" across the color TV screen). If you are not familiar with color TV principles, your attention is invited to the author's best-selling *Handbook of Simplified Television Service* (Englewood Cliffs, N.J.: Prentice-Hall, Inc., 1977).

A vectorscope permits the phase relationship (and amplitude) of the 10 color signals to be displayed as a *single pattern* on the oscilloscope screen. The vectorscope monitors the color signals directly at the color-gun inputs (grids or cathodes) of the color picture tube. This makes the vectorscope adaptable to any type of color circuit (any type of color demodulation, vacuum tube, solid-state, etc.). By comparing the vectorscope display against that of an ideal display (for phase relationship, amplitude, general appearance, etc.), the condition of the receiver color circuits (from antenna to picture tube) can be analyzed.

A conventional oscilloscope can be used as a vectorscope. This requires special connections to the horizontal and vertical deflection system, as described in Chapter 13. Oscilloscopes with vectorscope provisions need no special connections, although some provide a special cable and require special settings of the controls. Similarly, there are vectorscopes used in TV broadcast work that have special deflection circuits to provide the vector display.

1-8-1. How a Vector Display Is Obtained

A vector pattern is composed of two signals: one applied to the oscilloscope horizontal plates and the other to the vertical plates, as shown in Fig. 1-8. If two sine-wave signals are applied, and one is 90° out of phase but at the same frequency, the result is a circle as shown.

If the patterns formed by a standard 10-color-keyed rainbow generator are taken from the red and blue grids of a color picture tube, and these signals are applied to the vertical and horizontal deflection plates of an oscilloscope, the result is a pattern similar to that shown in Fig. 1-9. This pattern is generally referred to as a *vector pattern* (or possibly a *vectorscope pattern*).

The phase between the R-Y (red) and B-Y (blue) signals can readily be seen on the oscilloscope. (Oscilloscopes with built-in vectorscope provisions usually have a vector overlay, marked off in degrees, that can be placed over the screen, as shown in Fig. 1-10.) A complete circle represents 360°. The screen markings can be superimposed over the display to make phase meas-

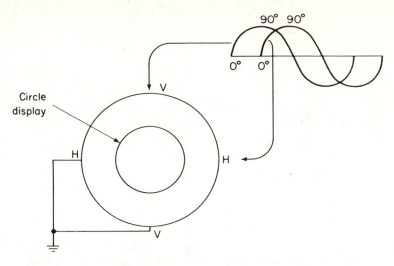

FIGURE 1-8 How a circular Lissajou pattern is formed (two sine waves 90° out of phase).

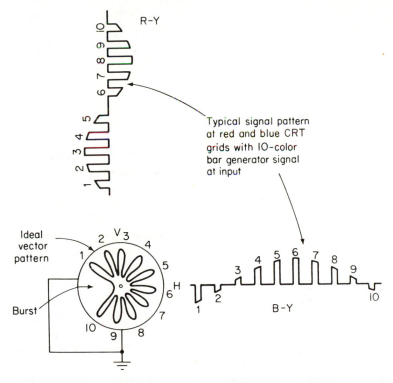

FIGURE 1-9 How a basic vector pattern is formed on a vectorscope.

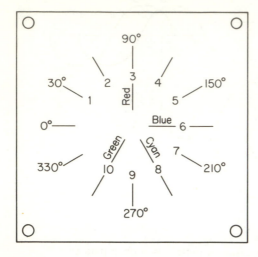

FIGURE 1-10 Typical vector overlay for vectorscope operation.

urements, in degrees. The length of the arm of the "petals" (patterns produced by each of the 10 color signals) may vary, depending on the signal. Similarly, the entire pattern rotates to a different position when the receiver "tint" or "hue" or other color control is adjusted. (Such controls change colors by slightly shifting the phase relationship of the receiver's color signals to the color burst transmitted by the TV broadcast station.)

1-8-2. Practical Vectorscope Connections

The horizontal and vertical inputs of a vectorscope are connected to the red and blue grids of the color picture tube as shown in Fig. 1–11. On some solid-state receivers, the color input is made at the color-gun cathodes instead of the grids. Either way, the green gun is not connected to present a vector display.

On some vectorscopes, the leads are permanently attached and are color-coded (red lead to red color gun, blue lead to blue gun). Other vectorscopes have connectors for the color-gun leads and a switch that converts the instrument from a conventional oscilloscope to a vectorscope. For some factory work, there are vectorscopes that represent a complete color analyzer system. Such instruments contain a color generator as well as the vectorscope.

The pattern of Fig. 1–9 represents an "ideal" vectorscope pattern, and is rarely (if ever) found in practical applications. Chapter 13 illustrates more realistic vector patterns. No matter what pattern is produced, interpretation is what counts. The problem of interpreting vectorscope patterns, as well as adjustments and troubleshooting with vectorscopes, are covered in Chapter 13.

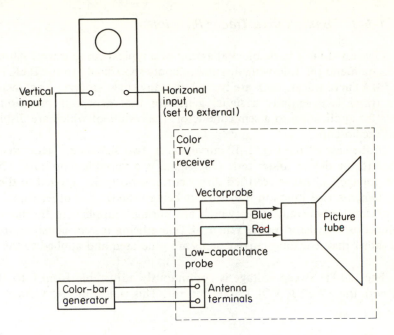

Vertical input

Horizontal input (set to external)

Color TV receiver

Vectorprobe

Blue
Red

Picture tube

Low-capacitance probe

Color-bar generator

Antenna terminals

FIGURE 1-11 Connecting a conventional oscilloscope as a vectorscope.

1-9. CURVE TRACERS

The most practical means of measuring transistor characteristics is to display the transistor characteristics as *curve traces* on an oscilloscope. Since the oscilloscope screen can be calibrated in voltage and current, the transistor characteristics can be read from the screen directly. If a number of curves are made with an oscilloscope, they can be compared with the curves drawn on transistor datasheets. (Some datasheet curves are reproductions of those obtained by tracing curves on an oscilloscope.)

There are a number of oscilloscopes (or oscilloscope adapters) manufactured specifically to display transistor and diode characteristic curves. The B&K Precision Semiconductor Curve Tracer is a typical unit. Some curve tracers are for only one type of transistor, whereas other instruments display the characteristics for many types (such as FETs, UJTs, SCRs, thyristors, PNPN devices) as well as diodes (signal, rectifying, zener, tunnel, step, etc.). The operating controls and basic operating procedures for typical curve tracers are described in Chapters 2 and 5, respectively. The procedures for using the curve tracer to display and test various solid-state devices are described in Chapter 9.

1-9-1. Basic Curve Tracer Principles

Figure 1–12 is a basic block diagram of a typical curve tracer. Although the diagram and the following discussion apply specifically to the B&K Precision 501A curve tracer, both are typical of many curve tracers. As shown, the curve tracer is essentially a signal generator that generates precision test signals for application to a semiconductor, the results of which are displayed on an oscilloscope.

In the case of the Fig. 1–12 curve tracer, two signals are generated and applied to the device under test. One signal is a variable-amplitude 120-Hz sweep voltage (full-wave rectified direct current) normally applied to the collector (when a two-junction transistor is being tested). The other signal consists of constant-current steps, which are normally applied to the base (for two-junction transistor tests). When FETs are being tested, constant-voltage steps rather than constant-current steps are generated and applied to the FET gate.

The 120-Hz sweep voltage is continuously adjustable from 0 to 100 V peak with the SWEEP VOLTAGE control. This pulsating d-c voltage is of

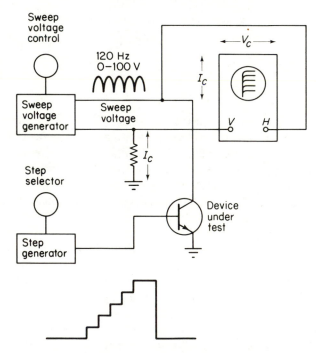

Constant–current steps for two-junction transistor; constant–voltage steps for FETS

FIGURE 1-12 Basic block diagram of typical curve tracer.

positive polarity when the POLARITY switch is set to NPN (N CHAN) and of negative polarity when the switch is set to PNP (P CHAN).

The step generator signal consists of five constant-current steps, plus a zero-current step, when two-junction transistors are being tested. The step selector switch offers 11 settings from 1 μA to 2 mA per step to match normal operating conditions for a full range of transistor types. When the polarity switch is in the NPN position, the steps are of positive polarity; in the PNP position, of negative polarity.

In the volts per step positions of the step selector switch, constant-voltage steps are generated for testing FETs. Five selections from 0.05 to 1 V per step are available. The polarity of the voltage steps is inverted in relation to the current steps. That is, the N CHAN position produces negative voltage steps, and the P CHAN position produces positive voltage steps (reverse-bias steps are required for testing FETs).

The sweep voltage that is applied to the collector of the transistor (or to the source of a FET) is also applied to the horizontal input of the oscilloscope. As the voltage increases from zero to maximum and returns to zero, a horizontal sweep is produced. A precision resistor in series with the sweep voltage source develops a voltage that is proportional to the resultant collector current. The voltage developed across this resistor is applied to the vertical input of the oscilloscope. Thus, vertical deflection represents collector current. Actually, four different precision resistors are included and selected by the vertical sensitivity switch on the curve tracer. This allows greater versatility for displaying a wide range of collector currents.

The steps of the step generator are synchronized with the pulses from the sweep generator. The base current remains at a fixed value while the sweep voltage increases from zero to maximum and returns to zero. The base current then steps to the next higher fixed value while the sweep voltage completes another cycle. The result is a family of six curves, one for each step, with each base current step increase producing a higher collector current. The display is a dynamic collector current I_C versus collector voltage (V_C) graph.

Most curve tracers are provided with a *graticule* or overlay that is placed on the oscilloscope screen. The graticule divides the display into 10 horizontal and vertical divisions for precise calibrated readings. When the curves show collector current versus collector voltage (for different values of base current), the change in collector current produced by one step of base current is proportional to the vertical distance between adjacent curves. This change is read directly from the graticule divisions. Which vertical line is chosen for the scale depends on what collector voltage is specified, because each vertical line corresponds to a particular collector voltage.

2

Basic
Operating Controls

This chapter deals with the operating controls of oscilloscopes. Because of the large variety of oscilloscopes it is impossible to discuss the controls of each make and model. As stated in the Preface, the functions of oscilloscope operating controls usually are fully covered in the related instruction manuals. Therefore, it is operating controls for typical oscilloscopes that are briefly described here.

All oscilloscopes have some operating controls in common. For example, most oscilloscopes have controls that position the trace vertically and horizontally. Not all controls, however, are found in any one oscilloscope. A simple, inexpensive oscilloscope will not have a single sweep control or operating mode where a single sweep can be displayed. Controls and connectors are not always found in the same location and in the same form on all oscilloscopes. On one instrument, the astigmatism control is a front-panel operating knob; on another, a side-panel screwdriver adjustment fulfills the same function.

Figures 2-1 and 2-2 show the front-panel operating controls of two representative oscilloscopes. As can be seen, both units have some controls in

FIGURE 2-1 B & K Precision Model 1535 Dual-Trace Oscilloscope (Courtesy of B & K—PRECISION Test Instruments Product Group of Dynascan Corporation, Chicago, Illinois).

FIGURE 2-2 B & K Precision Model 1460 Triggered Sweep Oscilloscope (Courtesy of B & K—PRECISION Test Instrument Product Group of Dynascan Corporation, Chicago, Illinois).

common (with the same, or different, names), but each oscilloscope has certain unique controls.

The control functions for each oscilloscope described are presented in the following sections.

2-1. B&K PRECISION MODEL 1535 DUAL-TRACE OSCILLOSCOPE

The B&K Precision Model 1535 Dual-Trace Oscilloscope is a solid-state portable oscilloscope designed to operate in a wide range of environmental conditions (see Fig. 2-1). The dual-channel dc to 35 MHz vertical system provides calibrated deflection factors from 2 mV/division. There are four modes of sync, including single sweep and fixed. The single-sweep mode allows a pushbutton to enable a single sweep, which begins at the next sync trigger. The fixed mode automatically fixes the sync level threshold at the center of the sync waveform. Five choices of sync coupling are available, including a sync separator for video signals.

The sync signal is selectable from five sources, including an alternate for dual-trace operation (channel A trace is triggered by the channel A input signal, and the channel B trace is triggered by the channel B input signal). Variable hold-off allows triggering to be inhibited beyond the sweep duration. A built-in signal delay line allows the leading edge of high-speed signals to be observed. UNCAL lights remind the operator that controls are not properly set for calibrated voltage or time measurements. For dual-trace operation, alternate or chop sweep modes can be selected automatically with sweep time or can be manually overridden. The instrument also provides for electrical trace rotation adjustable from the front panel, and a slotted bezel for mounting a standard oscilloscope camera.

The following paragraphs describe the function or operation of the controls and connectors, together with comments on how these controls relate to similar controls on other oscilloscopes.

MODE Switch. This five-position lever switch selects the basic operating modes of the oscilloscope. It (or a similar control) is found only on oscilloscopes having more than one mode of operation.

CH A.   Only the input signal to channel A is displayed as a single trace.

CH B.   Only the input signal to channel B is displayed as a single trace.

DUAL.   Dual-trace operation. Both the channel A and channel B input signals are displayed on two separate traces.

ADD.   The waveforms from channel A and channel B inputs are added and the sum is displayed as a single trace. When the channel B position is pulled (PULL INVERT), the waveform from channel B is subtracted from the channel A waveform and the difference is displayed as a single trace.

X-Y operation.   Channel A input signal produces vertical deflection (*Y*-axis). Channel B input signal produces horizontal deflection (*X*-axis).

TRACE ROTATION Control.   This control electrically rotates the trace to the horizontal position.

ASTIG Adjustment.   The astigmatism adjustment provides optimum spot roundness when used in conjunction with the focus and intensity controls. Very little readjustment of this control is required after initial adjustment.

Cathode-Ray Tube (CRT).   This is the screen on which the waveforms are viewed.

Scale.   This 8 × 10 cm graticule provides calibration marks for voltage (vertical) and time (horizontal) measurements. Illumination of the scale is fully adjustable. The engraved lines of the transparent viewing screen are brightened by edge-lighting the graticule. This provides a sharp reproduction of the lines when photographs are made from the screen, but does not produce an interfering glare.

FOCUS Control.   This control provides adjustment for a well-defined display. If a well-defined trace cannot be obtained with the focus control, it may be necessary to adjust the astigmatism control. To check for proper setting of the astigmatism adjustment, slowly turn the focus control through the optimum setting. If the astigmatism adjustment is correctly set, the vertical and horizontal portions of the trace will come into sharpest focus at the same position of the focus control. This setting of the astigmatism adjustment should be correct for any display. It may, however, be necessary to reset the focus control slightly when the intensity control is changed.

INTENSITY Control.   This control regulates the brightness of the display. The display can be adjusted from very bright to total darkness. The setting of the intensity control may also affect the correct focus of the display. Slight readjustment of the focus control may be necessary when the intensity level is changed.

To protect the CRT phosphor of any oscilloscope, do not turn the intensity control higher than necessary to provide a satisfactory display. Also, be careful that the intensity control is not set too high when changing from a fast to a slow sweep rate.

POWER SCAL ILLUM Control.   Fully counterclockwise rotation of this control (OFF position) turns off the oscilloscope. Clockwise rotation turns on the oscilloscope. Further clockwise rotation of the control increases the illumination level of the scale.

Pilot Light.   This light glows when the oscilloscope is turned on.

NORM-CHOP Switch.   This pushbutton switch operates in conjunction with the source switch to provide automatic or manual selection of alternate or chop method of dual-trace sweep generation.

With the source switch in the ALT position, the alternate sweep is selected regardless of sweep time and the NORM-CHOP switch has no effect.

With the source switch in any position except ALT and the NORM-CHOP switch in NORM (out position), the alternate sweep is automatically selected at all sweep times of 0.5 ms/cm and faster, and the chop sweep is automatically selected at all sweep times of 1 ms/cm and slower.

With the NORM-CHOP switch in CHOP (in position), the chop sweep is selected regardless of sweep time.

POSITION Control.   Rotation of this control adjusts the horizontal position of traces (both traces when operated in the dual trace mode). Also called *horizontal centering control, horizontal position control,* or *X-position control* on some oscilloscopes, this function moves the trace left and right to any desired horizontal position on the screen. The position control also has a push–pull function. When the position control is pulled out, the display is magnified five times. This display is normal when the position control is pushed in. The $5 \times$ magnification permits closer observation of a part of the signal. As shown in Fig. 2–3, the center divison of the unmagnified display is the portion visible on the screen in magnified form. Different portions of the display may be viewed by rotating the position control to bring the desired portion into the viewing area. A sweep magnifier control is usually found only on laboratory-type oscilloscopes.

Sweep Time UNCAL Indicator.   This lamp glows when the sweep time variable control is not set to CAL position, and reminds the user that time measurements are not calibrated.

SWEEP TIME/CM Switch.   This switch is the horizontal coarse sweep time selector. It selects calibrated sweep times of 0.1 μs/cm to 0.5 s/cm in 21 steps when the variable control is set to the CAL position (fully clockwise). On some shop-type oscilloscopes, this switch is called the *sweep range selector* or *coarse frequency control.* In such instruments, the switch positions are calibrated in terms of frequency rather than time (typically 10 to 100 Hz, 100 to 1000 Hz, 1 to 10 kHz, 10 to 100 kHz). In laboratory-type oscilloscopes, the sweep rates are expressed in time for convenience, since the time interval of the display is of greater importance than the sweep frequency

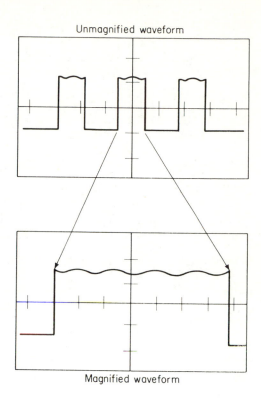

Unmagnified waveform

Magnified waveform

FIGURE 2-3 Operation of a sweep magnifier.

for most scientific measurements. If time is not given directly, it must be calculated from frequency, as is true for most older shop-type oscilloscopes.

Sweep Time VARIABLE Control. This control is the fine sweep time adjustment. In the extreme clockwise (CAL) position, the sweep time is calibrated as set by the SWEEP TIME/CM switch. In other positions, the variable control provides a continuously variable sweep rate. On shop-type oscilloscopes, the variable control is usually called the *sweep frequency control, fine frequency control,* or *frequency vernier,* and permits continuous variation of sweep frequency within any of the ranges provided by the sweep range selector.

HOLD-OFF Control. This control adjusts hold-off time (trigger inhibit period beyond sweep duration). It is usually set to the NORM position (fully counterclockwise) because no hold-off period is necessary. The hold-off control is useful when a complex series of pulses appears periodically, such as in Fig. 2–4a. Improper sync may produce a double image as in Fig. 2–4b. Such a display can be synchronized with the sweep time variable control, but this is impractical because time measurements are then uncalibrated. An alternative

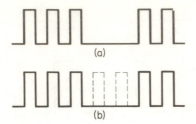

FIGURE 2-4 Typical use of the hold-off function: (a) with hold-off; (b) without hold-off.

method of synchronizing the display is with the hold-off control, which adjusts the duration of a period after the sweep in which triggering is inhibited. The sweep speed remains the same, but the triggering of the next sweep is "held off" for the duration selected by the hold-off control. At the MAX setting, the hold-off period is about 10 times greater than at the lowest setting.

EXT TRIG Jack. This jack is the input terminal for external trigger signal.

CAL 1-kHz O.1-V Terminal. This terminal provides a calibrated 1-kHz, 0.1-V peak-to-peak square-wave input signal. This signal is used for calibration of the vertical amplifier attenuators, and to check the frequency compensation of the probes used with the oscilloscope (Sec. 4–8).

RESET Button. When the triggering mode switch is in SINGLE, pushing the reset button initiates a single sweep that begins when the next sync trigger occurs.

READY/TRIG'D Indicator. In the single triggering mode, this indicator lights when the reset button is pressed and goes off when sweep is completed. In the norm, auto, and fix triggering modes, the indicator lights for the duration of the triggered sweep. The indicator also shows when the level control is properly set to obtain triggering.

LEVEL Control. Rotation of the triggering LEVEL control varies the triggering threshold level as illustrated in Fig. 2–5. In the + direction (clockwise), the triggering threshold shifts to a more positive value, and in the − direction (counterclockwise), the triggering threshold shifts to a more negative value. When the level control is centered, the threshold level is set at the approximate average of the signal used as the triggering source. The READY/TRIG'D indicator lights when the sweep is triggered, indicating that the triggering LEVEL control is within the proper range. If the TRIG'D indicator does not light at the center setting of the triggering level control, there

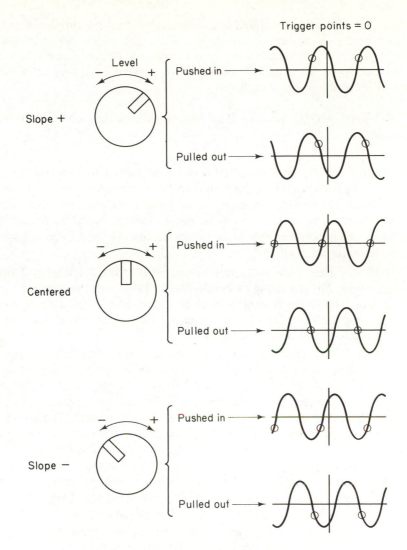

FIGURE 2-5 Function of the level control.

is insufficient sync signal for triggered operation, or one of the sync selection switches may be improperly set.

When the triggering level control is pushed in, the sync trigger is developed from the positive-going slope of the sync signal, and when the control is pulled out (PULL SLOPE NEG), the sync trigger is developed from the negative-going slope of the sync signal. The push–pull action of the level control also selects positive (in) or negative (out) polarity of composite video pulses when the coupling switch is in the video position. Figure 2–5 shows slope + and slope − triggering from a sine wave. However, any waveshape produces a trigger when the signal passes through the trigger threshold setting. Rotation of the level control adjusts the phase for which sweep triggering begins on sine-wave signals. For any type of triggering signal, the level control can be used to set an amplitude threhsold to prevent triggering by noise or undesired portions of the waveform.

Triggering MODE Switch. This four-position switch selects the triggering mode.

SINGLE. This position enables the reset switch for triggered single-sweep operation. When the signal to be displayed is not repetitive or varies in amplitude, shape, or time, a conventional repetitive display may produce an unstable presentation. The single-sweep mode is used to avoid this condition. The single sweep can also be used to photograph a nonrepetitive signal.

NORM. This is the most commonly used setting for triggered sweep operation. The triggering threshold is adjustable with the triggering level control. No sweep is generated in the absence of triggering signal or when the LEVEL CONTROL is set so that the threshold exceeds the amplitude of the triggering signal.

AUTO. This position selects automatic sweep operation where the sweep generator free-runs to generate a sweep without a trigger signal (this is called *recurrent sweep* operation on many oscilloscopes). In AUTO, the sweep generator automatically switches to triggered sweep operation if an acceptable trigger source signal is present. The auto position is handy when first setting up the scope to observe a waveform, since auto provides a sweep for waveform observation until other controls can be properly set. Auto sweep operation must be used for dc measurements and signals of such low amplitude that they do not trigger the sweep. Typically, signals that produce even 0.5 cm of vertical deflection are adequate for normal triggered sweep operation.

FIX. This mode is the same as the auto mode except that triggering always occurs at the center of the sync trigger waveform, regardless of the level control setting. This is very handy because it eliminates the need to constantly readjust the level control when making waveform measurements at several points which may be of different amplitude and waveshape. However, if triggering is desired at any point other than the center of the waveform, the norm or auto modes must be selected. Figure 2–6 shows the relationship between the trigger point and the trigger mode.

COUPLING Switch. This five-position level switch selects coupling for the sync trigger signal.

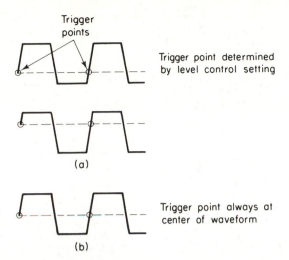

Trigger points

Trigger point determined by level control setting

(a)

Trigger point always at center of waveform

(b)

FIGURE 2-6 Relationship between trigger point and trigger mode: (a) norm or auto mode; (b) fix mode.

AC Position. This is the most commonly used position. AC position permits triggering from 10 Hz to over 35 MHz, and blocks any d-c component of the sync trigger signal.

LF Rej Position. In the LF Rej position, d-c signals are rejected, and signals below about 10 kHz are attenuated. Therefore, the sweep is triggered only by the higher-frequency components of the signal. This position is particularly useful for providing stable triggering if the trigger signal contains line-frequency components such as 60-Hz hum, as shown in Fig. 2–7a.

HF Rej Position. The HF Rej position attenuates signals above 100 kHz. D-c signals are rejected. This position is used to reduce high-frequency noise, and is especially useful to trigger on modulation rather than RF signals when observing amplitude-modulated waveforms.

VIDEO Position. This position is used primarily to view composite video signal in television circuits. In VIDEO, the sync trigger signal is routed through a sync separator circuit. The video position operates in conjunction with the SWEEP TIME/CM switch to automatically select horizontal or vertical sync pulses at appropriate sweep speeds for viewing each type of waveform. Vertical sync pulses (FRAME) are automatically selected at sweep speeds of 0.1 ms/cm and slower, since these sweep speeds are appropriate for viewing vertical frames of video. Horizontal sync pulses (LINE) are automatically selected at sweep speeds of 50 μs/cm and faster, since these sweep speeds are appropriate for viewing horizontal lines of video.

Without LF
rejection
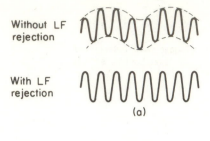

With LF
rejection

(a)

Without HF
rejection

With HF
rejection

(b)

FIGURE 2-7 Effects of LF and HF rejection on waveform displays. (a) Waveforms with low-frequency noise; (b) waveforms with high-frequency noise.

DC Position. This position permits triggering from dc to over 35 MHz. DC can be used to provide stable triggering with low-frequency signals which would be attenuated in the AC position, or with low-repetition rate signals. The level control can be adjusted to provide triggering at the desired d-c level on the waveforms. When using internal triggering, the setting of the vertical position controls may affect the d-c trigger level.

Z-AXIS INPUT Jack (mounted on rear panel). This connector is for intensity modulation of the CRT display (Sec. 1–5–3). The trace displayed on the screen may be intensity-modulated (*Z*-axis input) where frequency- or time-scale marks are required. A 5-V peak-to-peak or greater signal applied to the *Z*-axis input jack provides alternate brightness and blanking of the trace. A positive voltage input increases brightness. Figure 2–8 shows typical displays with *Z*-axis modulation.

Channel A and B POSITION Controls. These controls provide vertical position adjustment for the channel A and B traces. The controls become horizontal position controls when the mode switch is in the X-Y position. The controls are also push–pull, and select normal or inverted polarity of the channel A and B displays (display is inverted when the control is pulled).

Channel A and B UNCAL Indicators. These indicators glow when the A variable and B variable controls are not set to CAL positions. This reminds the user that the channel A or B measurements are not calibrated.

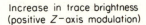

Increase in trace brightness
(positive Z-axis modulation)

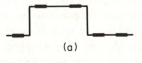

(a)

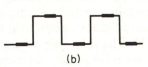

FIGURE 2-8 Typical (a) pulse and (b) square-wave displays with Z-axis modulation.

(b)

Channel A and B INPUT Jacks. These are the vertical input jacks for channels A and B. The jacks are used for external horizontal input when the mode switch is in the X-Y position.

Channel A and B AC-GND-DC Switches. These switches have three positions.

AC: blocks d-c component of input signal.

GND: opens the signal path and grounds the input of the vertical amplifier. This provides a zero-signal base line, the position of which can be used as a reference when performing d-c measurements.

DC: direct input of both a-c and d-c components of the input signal.

Channel A and B VOLTS/CM Switches. These switches are vertical attenuators for channel A and B. Both switches provide step adjustment of vertical sensitivity, and are calibrated in 12 steps, from 2 mV/cm to 10 V/cm (when the corresponding variable control is set to CAL position). The V/cm control adjusts horizontal sensitivity when the mode switch is in the X-Y position.

Channel A and B VARIABLE Controls. These controls provide fine control of vertical sensitivity. In the extreme clockwise (CAL) position, the vertical attenuator is calibrated (as determined by the corresponding V/cm switch). The variable controls become the fine horizontal gain control when the mode switch is in the X-Y position.

2-2. B&K PRECISION MODEL 1460 TRIGGERED SWEEP OSCILLOSCOPE

The B&K Precision Model 1460 solid-state oscilloscope is designed for both laboratory and service or shop work (see Fig. 2-2). The single-channel dc to 10-MHz vertical system provides calibrated deflection factors from 0.01 to 20

V/cm. The horizontal sweep time is 0.5 μs/cm to 0.5 s/cm, divided for 19 ranges, each providing means for fine adjustment. Both automatic and triggered sweep are available. Triggering may be internal, line, or external (2 V peak to peak), and may be positive or negative with a triggering range of 20 Hz to 10 MHz. The display may also be synchronized with television signals (TVV and TVH, as discussed in Chapter 13). Intensity (Z-axis) modulation, sweep magnification, a calibration voltage (1-kHz square wave of 5 V peak to peak), and vectorscope functions are provided.

The functions of many controls on the Model 1460 are similar to those on the Model 1535. The following paragraphs describe those controls on the Model 1460 that are substantially different from the Model 1535.

VOLTS/CM Switch.   This switch is the vertical attenuator, and provides coarse adjustment of vertical sensitivity. Vertical sensitivity is calibrated in 11 steps from 0.01 to 20 V/cm when the variable control is set to CAL position.

VARIABLE Control.   This control is the vertical attenuator adjustment, and provides fine control of vertical sensitivity. In the extreme clockwise (CAL) position, the vertical attenuator is calibrated (as determined by the V/cm switch).

DC BAL Adjustment.   This adjustment balances the vertical d-c amplifiers for trace centering.

SWEEP TIME/CM Switch.   This switch is the horizontal coarse sweep time selector, and selects calibrated sweep times of 0.5 μs/cm to 0.5 s/cm in 19 steps when the VARIABLE/HOR GAIN control is set to CAL. The switch selects proper sweep time for viewing television composite video waveforms in the TVH and TVV positions. The switch disables the internal sweep generator and displays external horizontal input in the EXT position.

VARIABLE/HOR GAIN Control.   This control is the fine sweep time adjustment (horizontal gain adjustment when the SWEEP TIME/CM switch is in the EXT position). In the extreme clockwise position (CAL), the sweep time is calibrated (as determined by the SWEEP TIME/CM switch).

V INPUT Jack.   This jack provides for input to the vertical amplifiers.

TRIG LEVEL Control.   This control is a sync level adjustment, and determines the point on the waveform slope where sweep starts. In the fully counterclockwise (auto) position, sweep is automatically synchronized to the average level of the waveform.

STABILITY Control.   This control provides for adjustment of sync stability.

EXT SYNC/HOR Jack.   This jack is the input terminal for external sync or external horizontal input.

TRIGGERING SLOPE Switch.   This switch selects sync polarity (+) or (−).

TRIGGERING SOURCE Switch.   This switch has three positions:

INT: waveform being observed is used a sync trigger.
EXT: signal at EXT SYNC/HOR jack is used as sync trigger.
LINE: power-line frequency sync (50/60 Hz).

SYNC Switch.   This switch has three positions:

NOR: normal sync.
TVH: syncs on horizontal components of composite video.
TVV: syncs on vertical components of composite video.

2-3. MISCELLANEOUS OPERATING CONTROLS

In addition to the controls of major importance described in Secs. 2-1 and 2-2, other controls must be adjusted for proper oscilloscope performance. On some oscilloscopes, these additional functions are adjusted by front-panel controls; on other instruments, the function is set by a screwdriver adjustment.

One front-panel control, found particularly on oscilloscopes used for television service, is a "phasing control." Adjustment of this control varies the horizontal sweep when it is being driven by the line voltage at the line frequency. The "phasing control" is especially important when the oscilloscope is used with a sweep generator (driven at line frequency) to observe a response pattern of a tuned circuit or amplifier. If there is a phase shift between the sweep generator and horizontal sweep, even though they are at the same frequency, a double pattern will appear, as shown in Fig. 2-9. This condition can be corrected by shifting the oscilloscope sweep drive signal phase.

FIGURE 2-9 Double RF response curve caused by improper setting of phasing control (no blanking used).

The following additional controls or adjustments are found in many oscilloscopes:

VOLTAGE REGULATION: sets the output voltage of regulated power supplies in oscilloscopes.

CALIBRATION VOLTAGE: sets the calibration voltage output.

SWEEP FREQUENCY: sets the horizontal sweep oscillator frequencies to match the sweep selector switch calibrations.

LINEARITY: sets linear horizontal and vertical deflection on each side of the screen center.

FREQUENCY COMPENSATION: sets amplifier and attenuator components (vertical and horizontal) for wideband response.

HUM BALANCE: cancels power supply hum.

2-4. B&K PRECISION MODEL 501A SEMICONDUCTOR CURVE TRACER

Figure 2–10 shows the operating controls for the B&K Precision Model 501A curve tracer discussed in Sec. 1–9. The following are descriptions of these con-

FIGURE 2-10 B & K Precision Model 501A Semiconductor Curve Tracer (Courtesy of B & K—PRECISION Test Instrument Product Group of Dynascan Corporation, Chicago, Illinois).

trols. The procedures for using the curve tracer controls are discussed in Sec. 5-6.

SWEEP VOLTAGE Control.   This control is a combination on–off switch and sweep voltage control. In the completely counterclockwise (AC OFF) position, power is removed from the curve tracer. Clockwise rotation of the control turns the curve tracer on, and sets the maximum value of the pulsating voltage at the collector (C) terminals. The sweep voltage is continuously adjustable from 0 to 100 V.

VERTICAL SENSITIVITY Switch.   This switch selects the vertical sensitivity of the oscilloscope so the collector current can easily be read. Four scales are provided. The scales equal 1, 2, 5, and 10 mA/division of the graticule scale when the oscilloscope has been calibrated.

H terminal: provides for connecting the curve tracer output to the horizontal input of the oscilloscope.

G terminal: provides a ground terminal for the oscilloscope.

V terminal: provides for connecting the curve tracer output to the vertical input of the oscilloscope.

STEP SELECTOR Switch.   This 17-position rotary switch selects current or voltage steps.

CURRENT PER STEP positions: select the base-current step values. The curve tracer automatically generates base current in five increasing steps. The switch offers 11 selections from 1 μA to 2 mA per step.

I_{DSS}/I_{CES} position: shorts gate-to-source terminal for measuring "zero volts" gate bias drain-to-source current and shorts base-to-emitter terminal for measuring collector–emitter leakage current with "zero volts" base bias.

VOLTS PER STEP positions (polarity inverted): select the gate voltage step value for testing FETs. The curve tracer automatically generates gate voltage in five increasing reverse bias steps and offers five selections, from 0.05 to 1 V per step.

POLARITY Switch.   This switch selects correct polarity for testing NPN or PNP transistors, and N-channel or P-channel FETs.

SOCKET Switch.   This switch selects left or right sockets and terminals for test.

LEFT and RIGHT E, S Terminals.   These terminals connect to the emitter or source of the semiconductor to be tested.

LEFT and RIGHT B, G Terminals.   These terminals connect to the base or gate of the semiconductor to be tested.

LEFT and RIGHT C, D Terminals.   These terminals connect to the collector or drain of semiconductor to be tested.

LEFT and RIGHT Sockets.   These are test sockets for plug-in transistors or FETs.

3

Oscilloscope
Specifications
and Performance

This chapter discusses oscilloscope characteristics. Although this text is not intended as a "theory book" or an oscilloscope circuits book, it assumes that a working knowledge of oscilloscope characteristics will enable you to obtain maximum benefit from your instrument. The following discussion is also intended to explain many of the technical terms used to describe oscilloscopes.

3-1. SWEEPS AND SCALES

In most cases, oscilloscopes have built-in sawtooth sweep generators which produce constant-speed horizontal beam deflection. In early oscilloscopes, the generators ran continuously. Horizontal calibration was based on their repetition frequency. In laboratory oscilloscopes, sweeps are usually calibrated in terms of a direct unit of time for a given distance of spot travel across the screen. This accounts for the term *time base*.

An oscilloscope with the widest range of sweeps is usually the most versatile. The primary usefulness of an oscilloscope, however, is as a high-speed device. Fastest sweeps are usually considered adequate if they can display *one cycle of the upper passband frequency across the full horizontal scale*.

When accurate *rise-time* measurements must be made which require the fastest sweep, a useful figure of merit for the adequacy of that sweep is

$$M = \frac{T_R}{T_D}$$

where M = figure of merit
  T_R = vertical system (usually amplifier) rise time
  T_D = time per division of the fastest sweep

Figures of merit greater than 1 are seldom found in oscilloscopes having rise times less than approximately 30 ns. Figures of merit greater than about 6 exceed the ideal and offer no further advantage. Accurate rise-time measurements should not be attempted when the rise time of a step signal exceeds the vertical deflection system rise time.

Time-base accuracy is usually specified in terms of the permissible full-scale sweep timing error for any calibrated sweep. That is, an accuracy of 3% would mean that the actual full-scale period of any sweep should not be more than 3% greater or less than indicated.

Magnified sweeps may have poorer accuracy ratings than unmagnified sweeps, since magnification is usually achieved by reducing amplifier feed-back.

Portions of sweeps may be magnified by increasing the gain of the horizontal amplifier, allowing either or both ends of the sweep to go off-screen, and positioning the display so that the desired portion is on screen. This method delays the presention of a sweep portion. Another method is generating suitably delayed sweep triggering signals so that the fast sweeps may be triggered just before the moments when the signal to be examined occurs. This second method delays actual generation of the displayed sweep; it is usually the preferred method, since it provides better long-term accuracy of the displayed time base, eliminates jitter, and provides greater time-interval measurements.

3-2. RISE TIME AND HIGH-FREQUENCY RESPONSE

Rise time is the more important specification for faster oscilloscopes; and passband (bandwidth), the more frequently used specification for slower oscilloscopes. The product of rise time and frequency response should produce a factor whose value lies between 0.33 and 0.35, when transient response is optimum.

NOTE

Transient response is the faithfulness with which a deflection system displays fast-rising step signals. The most common transient response distortions are overshoot, ringing, and reflections from impedance discontinuities in the vertical signal delay line. These forms of distortion make "clean" step signals appear to have spikes, squiggles, or bumps when they actually do not. These forms of distortion also make "unclean" signals appear worse.

In an example of combining rise time and frequency response to determine optimum transient response, the product of 0.023 μs rise time (0.023 × 10^{-6} s) and 15 MHz (15 × 10^6 Hz) equals 0.345. Factors less than 0.35 probably indicate overshoot in excess of 2%; factors larger than 0.4 probably indicate overshoot in excess of 5%. Ideally, oscilloscopes should have a vertical system capable of rising in about one-fifth the time that it takes the fastest step signal to rise.

Signal rise time can be calculated to a close approximation by the equation

$$T_S = \quad T_I^2 - T_A^2$$

where T_S = signal rise time
       T_I = indicated rise time
       T_A = vertical system (usually amplifier) rise time

The accuracy of such calculations falls off sharply for signals that rise faster than the oscilloscope amplifier.

3-3.   DISPLAY GEOMETRY

Here are some common limitations and application pitfalls that apply to almost all oscilloscopes. Some easily made performance checks are also included.

3-3-1.   Trace and Scale Alignment

A horizontal trace should coincide with the horizontal scale markings on the graticule. Misalignment, as shown in Fig. 3-1, usually indicates a need to reorient the CRT or the scale, but may be caused by inadequate CRT shielding or the presence of a strong magnetic field. CRTs operated with low-accelerating potentials are most susceptible. Even the earth's magnetic field may alter trace alignment.

3-3-2.   CRT Deflection Perpendicularity

The horizontal plates should deflect the beam in a direction perpendicular to that of the vertical deflection plates. A small deviation would be less than 1° near the center of the screen. A 1° error is a displacement from perpendicular of 1mm in 5.7 cm. Axis inaccuracy will affect the whole display geometry.

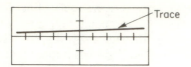

FIGURE 3-1 Example of misalignment between trace and scale.

3-3-3. Trace Bowing

Beam deflection may deviate from a straight line when a trace appears near the outer limits of the useful screen area, as shown in Fig. 3-2. A CRT may be tested by using sets of horizontal or vertical lines or by manually positioning the beam rapidly back and forth, horizontally and vertically, near all four sides of the useful area. Bowing tolerance depends upon the CRT type and its operating voltages. A typical high-quality CRT will not deviate from parallel lines by more than 1 mm on the edges or by more than 0.5 mm at the top and bottom.

To minimize bowing, some systems have adjustable voltages on special electrodes in the CRT. Excessive bowing may be due to improper accelerating voltage, a poor CRT, or an appreciable difference between the d-c levels of the vertical and horizontal deflection plates (with spot centered).

3-3-4. Horizontal Nonlinearity

Equal changes in voltage on the horizontal plates should produce equal changes in spot position. Any CRT nonlinearity (as shown in Fig. 3-3) that exists cannot usually be compensated for. Other nonlinearities depend upon the quality and adjustment of the sweep generator and/or horizontal amplifier.

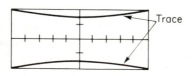

FIGURE 3-2 Example (exaggerated) of trace bowing.

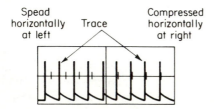

FIGURE 3-3 Example (exaggerated) of horizontal nonlinearity.

3-3-5. Amplitude Compression

Amplitude linearity of a d-c–coupled amplifier may be checked by observing any change of amplitude of a small signal while positioning the display through the useful scan area. For a-c–coupled deflection plates, compression can be checked by changing the input signal by measured increments and observing whether the displayed signal amplitude changes by exactly corresponding amounts.

3-4. POSITION DRIFT

High gain, d-c–coupled amplifiers are apt to drift appreciably. After turn-on, as much as an hour or more is sometimes required for the rate of drift to reduce to a minimum. After warm-up, the maximum amount of drift to be expected is often specified in terms of millivolts (or microvolts) per hour. The amount of position change that such a drift represents depends upon the deflection factor selected. For example, if the deflection factor is 1 mV/cm, and the drift specification is 1 mV/hr, the drift in any 1-hr period should not be greater than 1 cm. In most cases, the drift per hour is of little significance, since measurements generally take no more than 1 minute or so.

4

Oscilloscope Probes and Accessories

Like a test meter, an oscilloscope can perform most of its functions without accessories. One exception that applies to both instruments is a probe. In fact, many of the probes used with test meters can be used with oscilloscopes. To get the full benefit from a probe, the user needs a working knowledge of how the probe performs its function, even though the actual operating procedure is quite simple. Therefore, this chapter describes the probes most commonly used with modern oscilloscopes. The chapter also describes a few other accessories that can be used effectively with both shop-type and laboratory oscilloscopes.

4-1. THE BASIC OSCILLOSCOPE PROBE

In its simplest form, the oscilloscope probe resembles the "test prod" of a meter. It is essentially a thin metal prod connected to the oscilloscope input terminal through an insulated flexible lead. All but a small tip of the rod is covered with an insulated handle so that the probe can be connected to any point of a circuit without touching nearby circuit parts. Sometimes the probe tip is provided with an alligator clip so that it is not necessary to hold the probe at the circuit point.

Such probes work well on circuits carrying dc and audio-frequency ac. If, however, the ac is at a high frequency, or if the gain of the oscilloscope amplifier is high, it may be necessary to use a special *low-capacitance* probe. Hand capacitance in a simple probe can cause pickup of hum, particularly if the oscilloscope amplifier gain is set high. This can be offset by shielding in a low-capacitance probe. More important, the input impedance of the oscilloscope is connected directly to the circuit under test by a simple probe and may change the circuit operation. The low-capacitance probe contains a series capacitor and resistor which increase the oscilloscope impedance.

4–2. LOW-CAPACITANCE PROBES

The basic circuit of a low-capacitance probe is shown in Fig. 4–1. The series resistance R_1 and capacitance C_1, as well as the parallel or shunt resistance R_2, are surrounded by a shielded handle. The values of R_1 and C_1 are preset at the factory by screwdriver adjustment.

In most low-capacitance probes, the values of R_1 and R_2 are selected so that they form a 10:1 voltage divider between the circuit under test and the oscilloscope input. Hence, the operator must remember that voltage indications will be one-tenth of the actual value when the probe is used.

The capacitance value of C_1, in combination with the values of R_1 and R_2, also provides a capacitance reduction from approximately 3:1 to 11:1. R_1 and C_1 are usually factory adjusted and should not be disturbed unless recalibration is required. In that event, calibration should be performed as described in the probe or oscilloscope instruction manual.

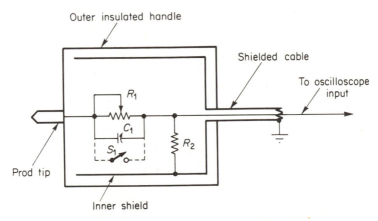

FIGURE 4–1 Typical low-capacitance probe circuit.

There are probes that combine the feature of low capacitance and the simple probes described in Sec. 4-1. In such probes, a switch (shows as S_1 in dashed form on Fig. 4-1) is used to short both C_1 and R_1 when a direct input is required. With S_1 open, both C_1 and R_1 are connected in series with the input, and the probe provides the low-capacitance feature.

4-3. EMITTER-FOLLOWER PROBES

The emitter-follower probe also provides a means of connecting into a circuit without disturbing circuit operation. The basic circuit of an emitter-follower probe is shown in Fig. 4-2. The circuit is essentially an emitter-follower stage within a shielded handle. The circuit probe tip is connected to the base input; the emitter output is connected to the oscilloscope input. The base input resistor R_1 is of high resistance (usually several megohms); the emitter load resistor R_2 is of the same value as the oscilloscope input impedance. Therefore, the circuit under test sees a high impedance that does not disturb the circuit, whereas the oscilloscope sees a matched impedance. Such a circuit has one disadvantage: the emitter-follower stage requires external power. For that reason, the emitter-follower probe is used only in special applications.

4-4. RESISTANCE-TYPE VOLTAGE-DIVIDER PROBES

A resistance-type voltage-divider probe is used when the primary concern is reduction of voltage. The resistance-type probe, shown in Fig. 4-3, is similar to the low-capacitance probe described in Sec. 4-2, except that the frequency-

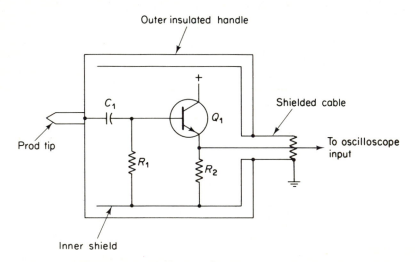

FIGURE 4-2 Typical emitter-follower probe circuit.

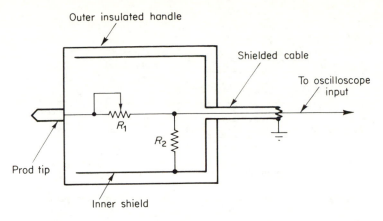

Outer insulated handle

Shielded cable

To oscilloscope input

R_1

R_2

Prod tip

Inner shield

FIGURE 4-3   Typical resistance-type voltage-divider probe circuit.

compensating capacitor is omitted. Usually, the straight resistance-type probe is used when a voltage reduction of 100:1, or greater, is required, and when a flat frequency response is of no particular concern.

As shown in Fig. 4–3, the values of R_1 and R_2 are selected to provide the necessary voltage division and to match the input impedance of the oscilloscope. Resistor R_1 is usually made variable so that an exact voltage division can be obtained.

Because of their voltage reduction capabilities, resistance-type probes are often known as *high-voltage probes*. Some resistance-type probes are capable of measuring potentials at or near 40 kV (with a 1000:1 voltage reduction).

4-5.   CAPACITANCE-TYPE VOLTAGE-DIVIDER PROBES

In certain isolated cases, the resistance-type voltage-divider probes described in Sec. 4–4 are not suitable for measurement of high voltages because stray conduction paths are set up by the resistors. A capacitance-type probe, shown in Fig. 4–4, can be used in those cases.

In such capacitance probes, the values of C_1 and C_2 are selected to provide the necessary voltage division and to match the input capacitance of the oscilloscope. Capacitor C_1 is usually made variable so that an exact voltage division can be obtained.

4-6.   RADIO-FREQUENCY PROBES

When the signals to be observed by an oscilloscope are at radio frequencies and are beyond the frequency capabilities of the oscilloscope amplifiers, a

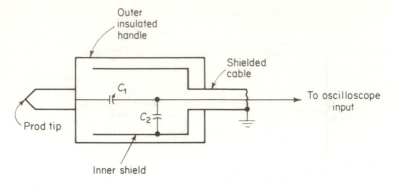

FIGURE 4-4 Typical capacitance-type voltage-divider probe circuit.

radio-frequency or RF probe is required. Such probes convert (rectify) the RF signals into a d-c output voltage which is equal (almost) to the peak RF voltage. The dc output of the probe is then applied to the oscilloscope input and is displayed on the screen in the normal manner. In most RF probes, the oscilloscope deflection is read as peak RF voltage. The probe circuit can be modified so that the probe dc output corresponds to the rms value of the RF signal voltage.

The basic circuit of a radio-frequency probe is shown in Fig. 4–5. Capacitor C_1 is a high-capacitance, d-c blocking capacitor used to protect diode CR_1. Usually, a germanium diode is used for CR_1, which rectifies the RF voltage and produces a d-c output voltage across R_1. This d-c voltage is equal to the peak RF voltage, less whatever forward drop exists across the diode CR_1. When it is desired to produce a d-c output voltage equal to the rms of the RF voltage, a series-dropping resistor (shown in dashed form on Fig. 4–5 as R_2) is added to the circuit. Resistor R_2 drops the d-c output voltage to a level that equals 0.707 of the peak RF value.

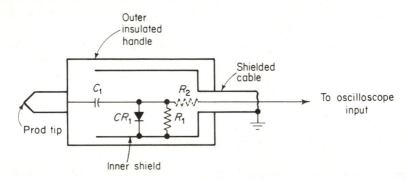

FIGURE 4-5 Typical radio-frequency probe circuit.

4-7. DEMODULATOR PROBES

The circuit of a demodulator probe (Fig. 4–6) is essentially like that of the RF probe described in Sec. 4–6. The circuit values and the basic function are somewhat different.

The prime purpose of a demodulator probe is demodulating an amplitude-modulated signal and converting the modulation envelope (low-frequency component) into a d-c output voltage.

The basic circuit of a demodulator probe is shown in Fig. 4–6. Here, capacitor C_1 is a low-capacitance, d-c blocking capacitor. (In the RF probe, a high capacitance value is required for C_1 to ensure that the diode operates at the peak of the RF signal. This is not required for a demodulator probe.) Germanium diode CR_1 demodulates (or detects) the amplitude-modulated signal and produces a voltage across load resistor R_1. This voltage is pulsating dc, proportional in amplitude to the modulating voltage, and it has the same approximate waveform and frequency as the modulating voltage. Resistor R_2 is used primarily for isolation between the circuit under test and the oscilloscope input.

4-8. USING PROBES EFFECTIVELY

The proper use of probes is important when reliable measurements are to be made. The following notes summarize some of the more significant considerations in the use and adjustment of probes.

4-8-1. Probe Compensation

The capacitors that compensate for excessive attenuation of high-frequency signal components (through the probe resistance dividers) affect the entire frequency range from some midband point upward. These capacitors

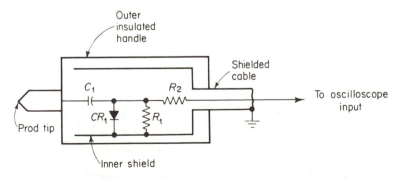

FIGURE 4-6 Typical demodulator probe circuit.

must be adjusted so that the higher-frequency components will be attenuated by the same amount as low frequencies and dc. Adjustment of the compensating capacitors is a factory job and should be accomplished using the proper test equipment. It is possible, however, to check adjustment of the probe-compensating capacitors using a square-wave signal source. First apply the square-wave signals directly to the oscilloscope input; then apply the same signals through the probe and note any change in pattern. There should be no change (except for a reduction of the amplitude) in a properly adjusted probe. Figure 4–7 shows typical square-wave displays with the probe properly compensated, undercompensated (high frequencies underemphasized) and overcompensated (high frequencies overemphasized).

Proper adjustment of probes is often neglected, especially when probes are used interchangeably with oscilloscopes having different input characteristics. It is recommended that any probe be checked with square-wave signals before use.

Another problem related to probe compensation is that input *capacitance* of the oscilloscope may change with age, or when the oscilloscope input tube is changed. Then all the compensated dividers (vertical input step attenuator) ahead of the input tube will be improperly adjusted. Readjustment of the probe will not correct for the change needed by the other dividers

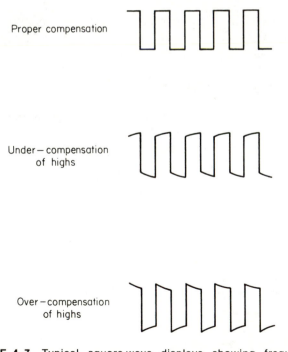

Proper compensation

Under–compensation of highs

Over–compensation of highs

FIGURE 4–7 Typical square-wave displays showing frequency compensation of probes.

of the step attenuator. The oscilloscope circuits must be adjusted instead. Also, when compensated dividers are cascaded in step attenuators, proper value of the resistors is especially important. Otherwise, proper compensation may not be possible at some attenuation ratios.

4-8-2. Circuit Loading

Connection of an oscilloscope to a circuit may alter the amplitude or waveform at the point of connection. To prevent this, the impedance of the circuit being measured must be a small fraction of the input impedance of the oscilloscope. When a probe is used, the probe's impedance determines the amount of circuit loading. The ratio of the two impedances represents the amount of probable error. For example, a ratio of 1:100 will account for about a 1% error; a ratio of 1:10, for about a 9% error. Remember that the input impedance is not the same at all frequencies but continues to diminish at higher frequencies because of input capacitance. Even at audio frequencies, the change in impedance may be significant. When using a shielded cable, the additional capacitance of the cable should be recognized when not terminated at one end in its characteristic impedance.

The reduction of resistive loading due to probes may be as much as the attenuation ratio of the probe, but capacitive loading will not be reduced to the same extent, because of the additional capacitance of the probe cable. A typical 5:1 attenuator probe may be able to reduce capacitive loading somewhat better than 2:1. A 50:1 attenuator probe may reduce capacitive loading by about 10:1. Beyond this point, little improvement can be effected because of the stray capacitance at the probe tip.

4-8-3. Practical Use of Probes

Whether a particular probe connection is disturbing a circuit can be judged by attaching and detaching another connection of similar kind (such as another probe) and observing any difference in the oscilloscope display.

Long probes should be restricted to the measurement of relatively slow changing signals. The same is true for long ground leads.

The ground lead should be connected to a point where no hum or high-frequency signal components exist in the ground path between that point and the signal pick-off point.

Reliable measurements involving frequency components higher than 10 MHz require probes with special inner conductors.

A 10-Ω resistor at the tip of a probe may prevent ringing of the ground lead when the probe is connected to very low impedance signal sources having very high frequency components.

Resistive loading may be eliminated entirely by using a small (0.002 μF or larger) capacitor in series with the probe tip, at the sacrifice of some low-frequency response.

Avoid applying more than the rated peak voltage to a probe. Using a high-voltage coupling capacitor between a probe tip and a very high d-c level may not always prevent probe burnout, since the capacitor must charge and discharge *through the probe*. If care is taken to charge and discharge the blocking capacitor through a path that shunts the probe, the technique can be successful. A recommended procedure is permanently to attach the blocking capacitor to the probe tip and to ground the junction of the capacitor and the tip whenever the capacitor is being charged or discharged.

Check for proper probe compensation whenever changing a probe or when making an important measurement.

4-9. ELECTRONIC SWITCH

In many applications, it is convenient to display two signals simultaneously. (Phase measurement by the dual-trace method described in Chapter 7 is such an application.) Many laboratory-type oscilloscopes have a dual-trace provision, where two signals can be applied directly to separate oscilloscope inputs and will be displayed simultaneously. The same type of operation can be accomplished with a shop-type oscilloscope when an electronic switch, or "chopper," is used at the input. The electronic switch acts as two gates, one for each signal, which open and close on alternate half-cycles at a predetermined frequency. One gate is open while the opposite gate is closed. The output of both gates is applied to the oscilloscope input. Therefore, both signals are displayed on the oscilloscope. In practice, the gate-switching frequency must be much higher than either signal frequency.

The basic circuit of a typical electronic switch is shown in Fig. 4–8. Each signal is applied to a separate gain control and gate stage. The gate stages are alternately biased to cutoff by square-wave signals from the square-wave generator. Therefore, only one gate stage is in a condition to pass its signal at any given time. The output of both gates is applied directly to the oscilloscope input.

Figure 4–8 also shows some typical dual-trace displays. Actually, each trace is composed of a tiny bit of the corresponding signal. Since the switching rate is fast, the trace appears as a solid line. Most electronic switches are provided with some form of positioning control so that the traces can be superimposed or separated, whichever is more convenient. (The positioning control is shown as R_3 in Fig. 4–8.) Also, so that the amplitude (or height) of both signals can be made to appear approximately the same on the display despite an actual difference in signal strength, the electronic switch is provided with separate gain controls (R_1 and R_2).

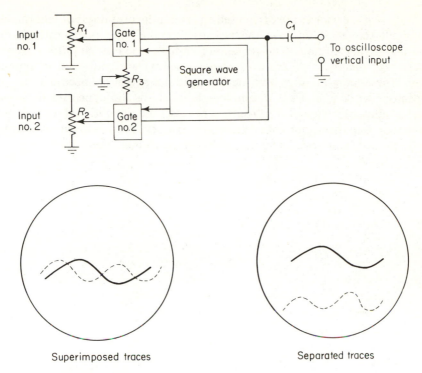

Superimposed traces   Separated traces

FIGURE 4-8 Electronic switch circuit and typical dual-trace displays.

4-10. OTHER ACCESSORIES

Many accessories besides probes and an electronic switch are available for use with oscilloscopes. Some of these are highly specialized units, such as diode or transistor switching-time testers. Other accessories are essentially extensions of the oscilloscopes, such as wideband amplifiers and time-base units with extended ranges. Still other oscilloscope accessories are basically test instruments that can be used on their own, such as square-wave generators, pulse generators, power supplies, time-mark generators, calibrators, and spectrum analyzers.

The basic operating principles, as well as detailed operating procedures, for these specialized accessories are described in the instruction manuals for the accessories, or in the manuals for the oscilloscopes with which the accessories are to be used. Therefore, oscilloscope accessories are not covered in further detail here. When accessories are required to perform the procedures discussed elsewhere in this book, the operating procedures and test connections for the accessories are described in the relevant chapters.

Plug-in accessories are often used to extend the usefulness of laboratory-type oscilloscopes. In this case, the basic oscilloscope consists of a cathode-ray tube and power circuits. The remaining circuits (such as amplifiers, sweep generators, time-base generators, trigger circuits) are contained in plug-in units. These plug-in units can be changed to perform specific tests. In laboratory work it is usual to employ the basic oscilloscope and those accessory plug-ins required for routine or essential operation (vertical and horizontal amplifiers, horizontal sweep generator). Then, as new work areas are developed, additional plug-ins are obtained to meet their needs.

5

Basic Operating Procedures and Recording Methods

The basic operating procedures of the various types of oscilloscopes described in Chapters 2 and 3 are discussed in this chapter; it also covers the basic methods for recording oscilloscope displays. Because of the great variety of oscilloscopes available, it is impossible to discuss the procedures for each make and model. Instead, typical units are described.

Later chapters describe how to use each type of oscilloscope to perform specific tests, alignments, and adjustments. There you will find instructions, such as "place oscilloscope in operation" or "set calibrating voltage to desired value." It is assumed that you will become familiar with your particular equipment. This is absolutely essential, since the procedures given in this chapter can be used only as a general guide to operating oscilloscope controls. You must understand each and every control on your particular equipment in order to follow the instructions of later chapters. No amount of textbook instruction will make you an expert in operating oscilloscopes; it takes actual practice. A thorough study of this chapter and a study of the controls on your particular instrument will put you on the right track.

NOTE

It is recommended that you establish a routine operating procedure, or sequence of operation, for each oscilloscope in the shop or laboratory. This will save time and will familiarize you with the capabilities and limitations of your particular equipment, thus eliminating false conclusions based on unknown operating characteristics.

5-1. PLACING AN OSCILLOSCOPE IN OPERATION

The first step in placing an oscilloscope in operation is reading the instruction manual for the particular oscilloscope. Although most instruction manuals are weak in applications data, they do describe turn-on, turn-off, and the logical sequence for operating controls; hence, the manual is of great help here, particularly if the operator is not familiar with the instrument.

After the manual's setup instructions have been digested, they can be compared with the following procedures. Remember that the procedures set down in this chapter are general or typical and applicable regardless of the test that is to be performed or the type of oscilloscope used. On the other hand, instruction-manual procedures apply to the specific instrument. Therefore, if there is a conflict between the manual procedures and the following instructions, follow the manual.

NOTE

Throughout the following chapters, the direction "place oscilloscope in operation (Chapter 5)" is used frequently. This direction refers to the following procedure, which has been put in this chapter to avoid repetition:

1. Set the power switch to off.
2. Set the internal recurrent sweep to off.
3. Set the focus, gain, intensity, and sync controls to their lowest position (usually full counterclockwise).
4. Set the sweep selector to external.
5. Set the vertical and horizontal position controls to their approximate midpoint.
6. Set the power switch to on. It is assumed that the power cord has been connected.
7. After a suitable warm-up period (as recommended by the instruc-

tion manual), adjust the intensity control until the trace spot appears on the screen. If a spot is not visible at any setting of the intensity control, the spot is probably off screen (unless the oscilloscope is defective). If necessary, use the vertical and horizontal position controls to bring the spot into view. *Always* use the lowest setting of the intensity control needed to see the spot. This will prevent burning the oscilloscope screen.

NOTE

The warm-up period of a d-c oscilloscope will be longer than that of an a-c oscilloscope, assuming that direct-coupled amplifiers are used in the d-c oscilloscope. Where a 1-min warm-up should be adequate for an a-c oscilloscope, most d-c oscilloscopes require 5 min. Also, many direct-coupled amplifiers will continue to drift for an hour or so after turn-on, as discussed in Chapter 3. Of course, if the oscilloscope uses all solid state circuits, warm-up time is almost instantaneous and drift is at a minimum.

8. Set the focus control for a sharp, fine dot.

9. Set the vertical and horizontal position controls to center the spot on the screen.

10. Set the sweep selector to internal. This should be the linear internal sweep, if more than one internal sweep is available.

11. Set the internal recurrent sweep to on. Set the sweep frequency to any frequency, or recurrent rate, higher than 100 Hz.

12. Adjust the horizontal gain control and check that the spot is expanded into a horizontal trace or line. The line length should be controllable by adjusting the horizontal gain control.

13. Return the horizontal gain control to zero (or its lowest setting). Set the internal recurrent sweep to off.

14. Set the vertical gain control to the approximate midpoint. Touch the vertical input with your finger. The stray signal pickup should cause the spot to be deflected vertically into a trace or line. Check that the line length is controllable by adjustment of the vertical gain control.

15. Return the vertical gain control to zero (or its lowest setting).

16. Set the internal recurrent sweep to on. Advance the horizontal gain control to expand the spot into a horizontal line.

17. If required, connect a probe to the vertical input.

18. The oscilloscope should now be ready for immediate use. Proceed with the detailed tests as described in later chapters.

NOTE

Depending upon the test to be performed, the oscilloscope may require calibration. Voltage and current calibration procedures are described in Chapter 6.

5-2. BASIC OPERATING PROCEDURE

Since an oscilloscope is essentially an item of test equipment, certain precautions must be observed during its operation. Many of the precautions are the same as those to be observed for a meter or signal generator; other precautions are unique to oscilloscopes. Some of the precautions are designed to prevent damage to the oscilloscope or the circuit under test; others are to prevent injury to the operator. The following precautions are divided into two groups: general safety precautions and oscilloscope operating precautions. Both should be studied thoroughly and then compared to any specific precautions called for in the oscilloscope's instruction manual.

5-2-1. General Safety Precautions

1. The metal cases of most oscilloscopes are connected to the ground of the internal circuit. For proper operation, the ground terminal of the oscilloscope should always be connected to the ground of the equipment under test. Make certain that the chassis of the equipment under test is not connected to either side of the a-c line (as is the case with some older a-c–d-c radio sets) or to any potential above ground. If there is any doubt, connect the equipment under test to the power line through an isolation transformer.

2. Remember, there is *always* danger inherent in testing electrical equipment that operates at hazardous voltages. Therefore, you should thoroughly familiarize yourself with the equipment under test before working on it, bearing in mind that high voltages may appear at unexpected points in defective equipment.

3. It is good practice to remove power before connecting test leads to high-voltage points. In fact, it is preferable to make all test connections with the power removed. If this is impractical, be especially careful to avoid accidental contact with equipment and other objects which can provide a ground. Working with one hand in your pocket and standing on a properly insulated floor lessens the danger of shock.

4. Filter capacitors may store a charge large enough to be hazardous. Therefore, discharge filter capacitors before attaching the test leads.

5. Remember that leads with broken insulation offer the additional hazard of high voltages appearing at exposed points along the leads. Check test leads for frayed or broken insulation before working with them.

6. To lessen the danger of accidental shock, disconnect test leads immediately after the test is completed.

7. Remember that the risk of severe shock is only one of the possible hazards. Even a minor shock can place the operator in danger of more serious risks, such as a bad fall or contact with a source of higher voltage.

8. The experienced operator continuously guards against injury and does not work on hazardous circuits unless another person is available to assist in case of accident.

5-2-2. Oscilloscope Operating Precautions

1. Even if you have had considerable experience with oscilloscopes, always study the instruction manual of *any* oscilloscope with which you are not familiar.

2. Use the procedures of Sec. 5–1 to place the oscilloscope in operation. It is good practice to go through the procedures each time that the oscilloscope is used. This is especially true when the oscilloscope is used by other persons. The operator cannot be certain that position, focus, and (especially) intensity controls are at safe positions, and the oscilloscope could be damaged by switching it on immediately.

3. As in the case of any cathode-ray-tube device (such as a TV receiver), the CRT spot should be kept moving on the screen. If the spot must remain in one position, keep the intensity control as low as possible.

4. Always use the *minimum intensity* necessary for good viewing.

5. If at all possible, avoid using an oscilloscope in direct sunlight, or in a brightly lighted room. This will permit a low-intensity setting. When the oscilloscope must be used in a bright light, use the viewing hood.

6. Make all measurements in the center area of the screen. Even if the CRT is flat, there is a chance of reading errors caused by distortion at the edges.

7. Use only shielded probes. Never allow your fingers to slip down to the metal probe tip when the probe is in contact with a "hot" circuit.

8. Avoid operating an oscilloscope in strong magnetic fields. Such fields can cause distortion of the display. Most good-quality oscilloscopes are well shielded against magnetic interference; however, the face of the CRT is still exposed and is subject to magnetic interference.

9. Most oscilloscopes have some maximum input voltage specified in the instruction manual. Do not exceed this maximum. Also, do not exceed the maximum line voltage, or use a different power frequency.

10. Avoid operating the oscilloscope with the shield or case removed. Besides the danger of exposing high-voltage circuits (several thousand volts are used on the CRT), there is the hazard of the CRT imploding and scattering glass at high velocity.

11.   Avoid vibration and mechanical shock. As is most electronic equipment, an oscilloscope is a delicate instrument.

12.   If an internal fan or blower is used, make sure that it is operating. Keep ventilation air filters clean.

13.   Do not attempt repair of an oscilloscope unless you are a qualified instrument technician. If you must adjust any internal controls, follow the instruction manual.

14.   Study the circuit under test before making any test connections. Try to match the capabilities of the oscilloscope to the circuit under test. For example, if the circuit under test has a range of measurements to be made (a-c, d-c, RF, pulse), you must use a wideband, d-c oscilloscope with a low-capacitance probe and possibly a demodulator probe. Do not try to measure 3-MHz signals with a 100-kHz bandwidth oscilloscope. On the other hand, it is wasteful to use a dual-trace 50-MHz laboratory oscilloscope to check out the audio sections of transistor radios.

NOTE

The most important oscilloscope operating precautions are summarized in Fig. 5–1.

5–3.   RECORDING AN OSCILLOSCOPE TRACE

In many applications, an oscilloscope trace must be recorded and not merely viewed. It is obviously much easier to measure and study a permanent record than an oscilloscope trace. Most recording is done with a Polaroid Land camera equipped with a special lens and a mounting frame attached to the oscilloscope. It is also possible to record with a conventional camera, with a moving film camera, or even by hand in some situations. The following sections briefly describe the basic recording methods. (Instruction manuals for oscilloscope cameras and accessories are quite detailed about operating details and applications.)

5–4.   OSCILLOSCOPE CAMERAS

Polaroid camera systems are most popular for oscilloscope work since they provide an immediate record with no wait for processing. One disadvantage of the early Polaroid system was that a positive print was obtained. This required special processing to obtain additional prints since no negative was available. One solution to this problem is to use a conventional camera back on the mounting frame. Graflok backs for cut film or film packs, or 120 and 620 roll film are used. However, negative-type Polaroid film has been

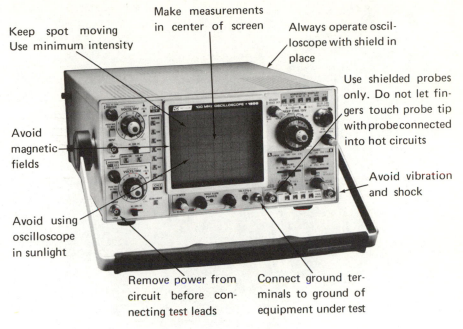

Keep spot moving
Use minimum intensity

Make measurements
in center of screen

Always operate oscil-
loscope with shield in
place

Use shielded probes
only. Do not let fin-
gers touch probe tip
with probe connected
into hot circuits

Avoid
magnetic
fields

Avoid vibration
and shock

Avoid using
oscilloscope
in sunlight

Remove power from
circuit before con-
necting test leads

Connect ground ter-
minals to ground of
equipment under test

FIGURE 5-1 Summary of oscilloscope operating precautions (Courtesy of B & K—PRECISION Test Instrument Product Group of Dynascan Corporation, Chicago, Illinois).

available for many years, so the use of conventional films is now limited to special applications.

Figure 5–2 is a photograph of a typical oscilloscope camera system. Figure 5–3 shows the basic operating principles of the system.

A special mount (bezel) is used to attach the camera to the oscilloscope. The bezel takes the place of the normal graticule cover on the oscilloscope. The optical system of the camera and attachment permits oscilloscope displays to be viewed and photographed simultaneously. Photographs are made directly from the oscilloscope screen so the image is not reversed. The viewed image is undistorted and is also not reversed. A lift-off mounting is used so that the camera can easily be mounted or removed. Swing-away hinges allow the camera to be swung out of the way when not in use. The viewing hood provides comfortable viewing with or without glasses. The rotating slide adapter allows any of the parfocal film-holding backs used with it to be locked in any of nine detented positions. All camera backs can also be rotated in 90° increments so that the long axis of the film will be parallel or perpendicular to the trace as desired. Several interchangeable lenses can be used with the camera when a rear casting is used. The parfocal backs allow photographs to be made on Polaroid Land or conventional film, in either sheet or roll film form.

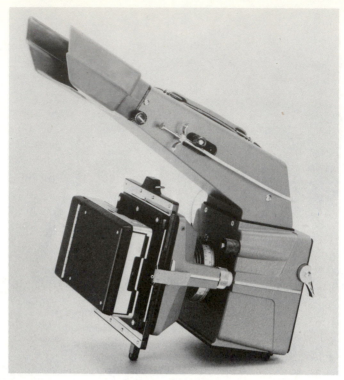

FIGURE 5-2 Tektronix oscilloscope camera system. (Courtesy of Tektronix.)

5-4-1. Optical System Operation

As shown in Fig. 5–3, the viewing system consists of a viewing hood and two mirrors. Light from the oscilloscope screen strikes the beam-splitting mirror where a portion of the light is transmitted to the camera lens and another portion is reflected to the second mirror. A virtual image acts as the object for the second mirror surface. The second mirror then forms a virtual image which is viewed by the observer. Owing to the 45° arrangement of the beam-splitting mirror, the observer views the oscilloscope display as though he or she were looking directly toward the oscilloscope screen on a line perpendicular to the screen. This orthogonal view is full size, but the image appears approximately 20 in. away. In all cases, the lens is considerably closer to the oscilloscope screen. The difference in the two distances produces a small amount of parallax between the viewed and photographed images. The small amount of parallax can usually be ignored.

5-4-2. Selecting a Camera Back

The camera back chosen depends primarily on the intended use for the photograph, how quickly you want the finished photograph, how large an area you wish to photograph, the magnification factor of the particular lens

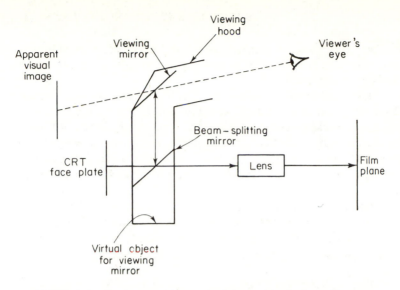

Apparent visual image

Viewing mirror

Viewing hood

Viewer's eye

CRT face plate

Beam – splitting mirror

Lens

Film plane

Virtual object for viewing mirror

FIGURE 5-3 Basic operating principles of typical oscilloscope camera system.

used, and the size of the negative desired. To obtain a negative from which a number of prints can be made, either type 55 P/N film (which comes in Polaroid Land 4 × 5 only) or conventional film is quite satisfactory. Both the Polaroid Land 4 × 5 film holder and the holders for conventional cut and roll film are used with the Graflok back in place.

With either Polaroid Land or conventional films, the size of the film used by the selected back must be at least as large as the image from the lens. This depends on the object-to-image ratio of the camera lens and on the size of the oscilloscope display. For example, the roll film back for 120 or 620 film would probably not be used with a 1:0.9 lens and a 10-cm-wide oscilloscope display because the image of the display is 9 cm wide and the long dimension of the film is only about 8.25 cm. Thus, at least 7.5 mm would be cut off the photograph. In practice, the film size should be at least 5 mm larger than the size of the image in order to allow for normal tolerances in the construction of the camera backs and for the position of the film in the back.

5-4-3. Writing Rate

The term *writing rate* is often used in oscilloscope photography. Writing rate is a figure of merit which roughly describes the ability of a particular camera system mounted on a particular oscilloscope to photograph fast-moving traces. The writing-rate figure expresses the maximum spot rate (usually in centimeters per microsecond) that can be photographed satisfactorily. The faster the oscilloscope spot moves, the dimmer the trace becomes because the electron beam strikes each point on the phosphor coating for a shorter period of time. A camera system and oscilloscope which have a high writing rate are required for low–repetition-rate displays at the fast oscilloscope sweep rates.

It is not practical to assign an absolute value of writing rate to any oscilloscope or camera, because so many variables are involved. Among the variables that must be considered are the type of film, the CRT accelerating potential, the camera optical arrangement, the object-to-image ratio of the camera lens, and development time of the film. It is possible to compare the effectiveness of two films by measuring their writing rate under the same conditions.

Although there is some relationship between ASA rating of film and writing rate, the ASA rating is not the final determining factor. It is safe to assume that a film with a very high ASA speed rating would probably have a higher maximum writing rate then a film with a lower ASA speed rating.

5-4-4. Selecting Film

For most oscilloscope work, Polaroid Land film is convenient. This film permits the picture to be seen very soon after taking it and makes it unnecessary to expose part or all of the film before developing it.

Table 5–1 gives a brief outline of the available Polaroid emulsions. The films recommended or films having equivalent characteristics may be used.

Table 5–2 lists conventional films available, and gives a brief description of their characteristics.

5-4-5. Selecting Cathode-Ray-Tube Phosphors

A great number of phosphor types are presently available to the purchaser of a cathode-ray oscilloscope. No one phosphor is best for all applications; each has advantages and disadvantages compared to the others. Of the many types of phosphors available, five are most commonly in use: the P1, P2, P7, P11, and P31. Other phosphor types are usually restricted to special applications.

For low sweep rate or repetitive-sweep applications where a high writing rate is not required, almost any type of phosphor is satisfactory. Selection of the CRT phosphor is important only for single-sweep or low-repetition-rate applications at the fast sweep rates. In low-repetition-rate applications at the fast sweep rates, use of the proper phosphor can mean the difference between getting a good photograph and getting none at all.

For photographing, the most important single characteristic of a phosphor probably is the color of its emitted light. A blue or violet fluorescence has the highest actinic value and thus is most suitable for photographic work. In general (all other factors being equal), the shorter the wavelength of the visible peak emitted light, the better the phosphor for photographic applications.

TABLE 5-1 Polaroid Land Film Types

Film type	ASA rating (approximate)	Picture size	Remarks
47	300	$3\frac{1}{4} \times 4\frac{1}{4}$	Panchromatic; paper print; roll film only; high-speed with medium contrast
107	3,000	$3\frac{1}{4} \times 4\frac{1}{4}$	Panchromatic; film packs only; paper print; similar to Type 47
410	10,000	$3\frac{1}{4} \times 4\frac{1}{4}$	Roll only; panchromatic type; paper print; extra-high-speed film good for extremely fast waveforms
46L	800	$3\frac{1}{4} \times 4\frac{1}{4}$	Roll only; yields positive transparency; medium contrast; high speed; panchromatic type
146L	125	$3\frac{1}{4} \times 4\frac{1}{4}$	Slower speed than type 46L; faster development time; roll only; positive transparency; high contrast
52	200	4×5	Sheet only; panchromatic type; yields paper print; good general-purpose film
55 P/N	50	4×5	Positive paper print and reproducible negative; sheet form; panchromatic type; high-resolution negative
57	300	4×5	Panchromatic type; sheet only; paper print; equivalent of type 47 in sheet form
48	75	$3\frac{1}{4} \times 4\frac{1}{4}$	Color film which yields a paper print; available in roll only; requires no coating
58	75	4×5	Color film which yields a paper print; available in sheet only; requires no coating
108	75	$3\frac{1}{4} \times 4\frac{1}{4}$	Color film which yields a paper print; available in film packs only; requires no coating

TABLE 5-2 Conventional Film Types

Manu-facturer	Film name	ASA rating	Remarks
Eastman Kodak	Tri-X	400	High speed, medium contrast; roll film
	RS Pan	650	Similar to Tri-X, in sheet form
	Royal-X Pan	1250	Ultra-fast roll film with low contrast
	Recording Royal-X Pan	1250	Same as Recording Royal-X Pan in sheet form
	Plus-X Pan	160	Medium-speed film with good contrast; both sheet and roll film
Agfa	Isopan Record	1000	High-speed film with low contrast; both sheet and roll film
Ansco	Super Hypan	400	Medium-speed film with medium contrast; both sheet and film

Most users of oscilloscopes are concerned not only with photographing the oscilloscope trace but in observing it directly as well. For such users, it is important to have a phosphor that gives good results in both types of applications. Frequently, the choice falls on a phosphor, such as P2 or P31, where the emitted light has large enough actinic value to give a good writing rate and also has sufficient persistence to permit easy viewing.

It has been observed that the P11 phosphor has the highest comparative writing rate of any common phosphor and is thus best for photographic work. The medium-short persistence of the phosphor is somewhat undesirable for general-purpose work, but the disadvantages of this are slight. Choose type P11 whenever the ultimate in photographic ability is required. Type P11 emits a blue light of medium-short duration.

Table 5-3 lists the characteristics of common phosphors.

TABLE 5-3   Common Phosphor Table

Phosphor type	Writing rate (% of P11, used as standard)	Relative brightness (%)	Color	
			Fluorescence	Phosphorescence
P1	35	150	Yellowish green	Yellowish green
P2	70	230	Bluish green	Green
P7	95	128	Blue-white	Yellowish green
P11	100	100	Purplish blue	Purplish blue
P31	75	390	Green	Green

5-4-6.   Photographic Recording Techniques

The following procedures can be used to obtain an exposure for both Polaroid and conventional film, when photographing *repetitive* oscilloscope traces.

1.   Position the external graticule, if any, for white lines.

NOTE

Graticule lighting should not be so high as to produce glare, but high enough to make the lines shine. This is true for almost all photographic recording, except where the exposure is very short. Medium graticule lighting for short exposures may not produce a sharp reproduction. Sometimes, this condition can be corrected by double exposure. The preferred technique is to expose the film first with the graticule only, then with the graticule and trace combined. As with most oscilloscope photography, the exposure times must be found by experiment. A dim, thin trace requires longer exposure but gives better reproduction. A bright trace can produce a "halo" or afterglow.

2. Mount the camera bezel on the oscilloscope.
3. Obtain the signal and adjust the controls for the desired display.
4. Attach the camera to the bezel and secure the camera against the oscilloscope.
5. Adjust the focus, astigmatism, and intensity controls for a sharp trace.
6. Set the aperture selector for the largest lens opening (smallest f-stop number) and carefully focus the camera on the trace or half-way between the trace and graticule.

NOTE

When using an external graticule and both clear trace and external graticule are desired, the camera should be focused halfway between the trace and graticule.

7. Set the intensity to midrange, graticule scale about three-quarters of full range, shutter speed to $\frac{1}{5}$ s, and aperture selector to f/5.6. These control settings should be reasonably close for film speeds of 400 ASA and a waveform frequency near 1 kHz. For film with a 3000 ASA rating and waveform frequency of 1 kHz, use a shutter speed of $\frac{1}{5}$ s and an aperture selector setting of f/4.5.

The following procedures can be used to obtain an exposure when photographing *single-sweep displays*.

NOTE

Single-sweep displays are formed when the oscilloscope spot sweeps across the screen only once. Actual exposure time is thus determined not by the shutter setting but by the duration of the sweep plus phosphor persistence, provided that the shutter is open sufficiently long. In one type of single-sweep photography, the graticule exposes the film for the time set by the shutter whereas the spot on the screen exposes the film for only the duration of the sweep. Therefore, it is not usually possible to adjust the trace and graticule for the same intensity and obtain good pictures, since the effective exposure times for the two are different. Success in obtaining good photographs of single-sweep displays will come only with experience. A few tips, however, may reduce the amount of experimenting required.

8. Select a shutter speed longer than the event to be photographed.

Alternatively, use the "time" or "bulb" camera position and hold the shutter open while manually triggering the sweep.

9. Use the highest possible intensity without causing defocus of the trace.

10. Where practical, use f-stops higher than f/4, if an external graticule is used. This will allow both trace and external graticule to be in focus.

NOTE

Remember that since the shutter speed has already been determined, the selection of lens opening will determine how well the trace photographs. In single-sweep applications, the camera settings must be made for trace intensity and duration. Graticule intensity cannot be used as a reference.

5-4-7. Conventional Cameras for Oscilloscope Recording

Conventional cameras can be used, with certain limitations, in place of Polaroid cameras for oscilloscope recording. The adapters used with Polaroid cameras can also be used with conventional film cameras, including 35 mm. The film must be processed in the normal manner.

It is also possible to use a conventional camera without an adapter, if several precautions are taken and if the inconvenience can be tolerated. Two basic problems are to be considered: focal length of the camera and ambient light.

Most cameras have a minimum focal length of 2 or 3 ft. This will produce a very small image on the negative, requiring blow-up of the print (with the usual distortion problems). This problem can be minimized using a camera fitted with a close-up lens. The ambient light problem can be overcome by using a cardboard tube between the oscilloscope screen and camera to exclude the light. The author has successfully used a twin-lens reflex with high-speed film (400 ASA minimum), focusing from a distance of about 12 in. through a hood or light shield. Any such arrangement is makeshift, however, and should be used as a temporary or emergency method.

5-4-8. Moving Film Cameras for Oscilloscope Recording

Moving film cameras are sometimes used in highly specialized oscilloscope work, such as studying lightning, noise, electrical breakdown, cosmic rays, metal fatigue, or any random occurrence. Moving film cameras are similar to motion picture cameras in that both have film which is drawn across the lens. In a moving film camera, the film is drawn continuously by an adjustable-speed motor, and there is no opening and closing of a shutter for each frame, as in a motion picture camera.

The moving film itself provides the horizontal sweep and the time base. Usually, the signal to be measured is applied to the oscilloscope vertical input in the normal manner, but the horizontal sweep is switched off. (Since operating a moving film camera for oscilloscope recording is so specialized, the instruction manuals provide a great amount of detail, which will not be repeated here.)

5-5. *HAND RECORDING OSCILLOSCOPE TRACES*

When a camera is not available, it is possible to "hand record" oscilloscope traces. Of course, such recording is limited to traces of long duration, which will remain stationary. Also, hand recording requires considerable skill and should be used only as a temporary or emergency technique.

The oscilloscope trace can be recorded on a transparent plastic overlay cut to the same size as the oscilloscope screen. Thin paper can also be used, but this requires that the intensity be advanced.

If the plastic overlay is used, the actual trace is made with a well-sharpened grease pencil, permitting the trace to be rubbed off and the overlay reused when the particular trace is no longer needed. Attach the overlay to the screen with two-sided adhesive tape, or hold firmly in place. It is often convenient to inscribe a graticule on the overlay, and to align this graticule with that of the oscilloscope screen. The plastic overlay should be thick enough so that it does not wrinkle, but not so thick that it produces parallax (0.01 in. is usually satisfactory).

If paper is used, the actual trace is made with a medium pencil, being careful not to scratch the screen. Also be careful not to advance the intensity to a point where the screen could be burned. It is often convenient to use graph paper, with a graticule ruled on one side. If possible, use a graph paper on which the divisions correspond to the oscilloscope screen divisions.

NOTE

The static electricity present on plastic or paper sheets may prove to be a problem when placed against the oscilloscope screen. The static electricity charge can distort the display. Discharge the paper or plastic before attaching it to the oscilloscope. This can be accomplished by touching the sheet to a good ground.

Types 47, 107, 410, 146L, 52, and 57 require 10 s of development time. Type 46L requires 2 min of development time. Type 55P/N requires 20 s of development time. Types 48, 58, and 108 require 50 to 60 s of development time.

5-6. PLACING A CURVE TRACER IN OPERATION

This section describes the basic operating procedures for the curve tracer described in Chapter 2. Although the procedures are for the B&K Precision curve tracer, they are typical for a great variety of curve tracers. Chapter 9 describes how curve tracers can be used to perform specific tests on solid-state devices.

5-6-1. Curve Tracer Setup

The first step in setting up a curve tracer for use with an oscilloscope is to read the instruction manuals for both the tracer and the oscilloscope. If there is a conflict between the manual procedures and the following instructions, follow the manuals.

1.   Connect the curve tracer and oscilloscope to suitable power sources.
2.   Set the step selector switch and vertical sensitivity switch to the fast-setup markers (triangular markers next to the controls). This provides a good starting point for all small-signal transistors and is low enough to prevent damage, even to very delicate transistors. It is recommended that these controls always be placed in this fast-setup reference position before connecting transistors for test.
3.   Turn the sweep voltage control clockwise to check that the power source is available. The pilot lamp should light when the sweep voltage control is turned clockwise from zero. Always return the sweep voltage control to zero before inserting a transistor for test.

5-6-2. Oscilloscope Setup and Calibration

Any 3-in. or larger general-purpose oscilloscope with external horizontal input is satisfactory for use with the curve tracer. The sweep circuits of the oscilloscope are not used, only the horizontal and vertical deflection circuits. The oscilloscope must have d-c inputs for both horizontal and vertical deflection, since a-c (capacitor-coupled) inputs produce trace shift and distort the curve display. Oscilloscope frequency bandwidth is not critical, but must be greater than 10 kHz (in both horizontal and vertical) for best results.

The oscilloscope must be calibrated with the curve tracer to get meaningful results. That is, the vertical divisions of the graticule must accurately represent the current through the device under test, and the horizontal divisions must accurately represent the sweep voltage applied to the device under test. In the case of the B&K Precision curve tracer, oscilloscope calibration is done with a calibrating source built into the tracer. The basic calibrating procedures are as follows.

5-6-3. Vertical Calibration

The curve tracer requires that the oscilloscope vertical gain be set for 1-V full-scale sensitivity (10 divisions of the special graticule). Once set, the display is accurately calibrated for all ranges of the vertical sensitivity switch on the curve tracer. An accurate (± 3%) 1-V peak-to-peak source is available at the base (B) jacks of the curve tracer when the step selector switch is placed in the 0.2 V per step position. The voltage steps from 0 to 1 V consist of five steps of 0.2 V each (six steps if zero volts is counted as a step), resulting in a series of six vertical dots on the oscilloscope screen. The vertical gain should be adjusted so one of the dots coincides with every second division, using the following procedure.

1. Place the special graticule over the oscilloscope screen. If necessary, cut the graticule to fit the oscilloscope.
2. Connect a test lead from the G (ground) jack of the curve tracer to the oscilloscope ground, as shown in Fig. 5–4.
3. Turn on the curve tracer and oscilloscope.
4. Set the oscilloscope for external horizontal operation, but do not apply a horizontal input at this time. The horizontal input may be grounded if desired.
5. Adjust the oscilloscope horizontal and vertical centering controls, if required, to place a dot in the approximate center of the screen. Do not allow a spot to burn the CRT.

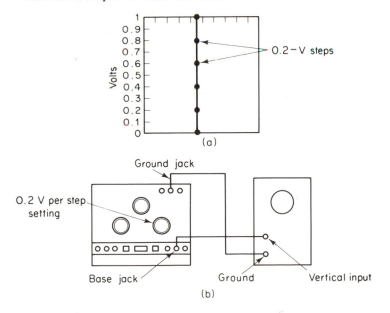

FIGURE 5-4 (a) Vertical calibration with (b) curve tracer.

6. Connect a test lead from one of the B (base) jacks of the curve tracer to the vertical input of the oscilloscope, as shown in Fig. 5–4. If the left B jack is used, the socket switch must be in the left position, and vice versa.

7. Set the step selector switch of the curve tracer to the 0.2 V per step position.

8. The display should resemble that shown in Fig. 5–4 (a series of six vertical dots, and possibly a light vertical trace between dots). Adjust the oscilloscope vertical gain so that the display fills the graticule from top to bottom exactly (10 divisions). Adjust the vertical centering as required.

9. This completes vertical calibration of the oscilloscope. Once the oscilloscope is calibrated, *do not* readjust the vertical gain. During operation, vertical positioning may be readjusted as required, but the display must be adjusted to the desired size by *controls on the curve tracer.*

5-6-4. Horizontal Calibration

A horizontal calibration of 1 V/division can be very accurately obtained using the B (base) jack output of the curve tracer when the step selector switch is in the 1 V per step position. The voltage steps from 0 to 5 V in five steps of 1 V each (six steps if zero volts is counted as the first step), resulting in a series of six horizontal dots on the oscilloscope screen, as shown in Fig. 5–5. The horizontal gain can be adjusted so that the dots coincide with five divisions of the graticule. This gives a full-scale value of 10 V, which is adequate for testing most small-signal transistors. Only slightly less accuracy is obtained by adjusting the horizontal gain for two dots per division (Fig. 5–5), thus giving a full-scale value of 20 V.

For higher test voltages, a less accurate ($\pm 15\%$) method of reading the horizontal voltage is available by connecting the sweep voltage output of the curve tracer (H jack) to the horizontal input of the oscilloscope. This method produces a *horizontal trace* as shown in Fig. 5–6, rather than a series of dots (Fig. 5–5). The maximum sweep voltage can be set for any desired convenient value, such as 50 or 100 V, as read from the sweep voltage dial. The horizontal gain of the oscilloscope can then be set for a full-scale horizontal trace (10 divisions of the graticule). Each division then equals 10% of the maximum sweep voltage.

5-6-5. Low-Voltage Horizontal Calibration

Adjust the horizontal gain using a low reference voltage as follows.

1. Complete the vertical calibration as described in Sec. 5–6–3.

2. Leave the oscilloscope set for external horizontal input, and the

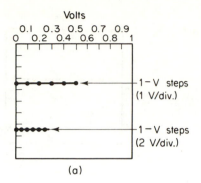

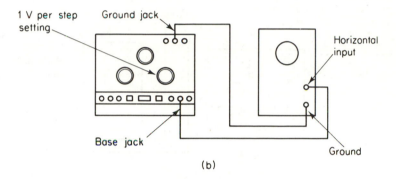

FIGURE 5-5 (a) Horizontal calibration with (b) curve tracer.

common ground connected between the curve tracer and the oscilloscope, as shown in Fig. 5-5.

3. Disconnect the vertical input from the oscilloscope. Ground the vertical input, if desired.

4. Connect a test lead from one of the B (base) jacks on the curve tracer to the horizontal input of the oscilloscope, as shown in Fig. 5-5. If the left B jack is used, the socket switch must be in the left position, and vice versa.

5. Set the step selector switch on the curve tracer to the 1 V per step position.

6. The display should resemble that shown in Fig. 5-5 (a series of six horizontal dots, and possibly a light horizontal trace between dots). Adjust the oscilloscope horizontal gain so that the display fills exactly five divisions of the graticule. Adjust the horizontal and vertical centering controls as required.

7. This completes horizontal calibration of the oscilloscope at 1 V/division. Once the oscilloscope is calibrated, *do not* readjust the horizontal gain. During operation, horizontal centering may be readjusted as required.

8. To calibrate the horizontal display for 2 V/division, follow the same

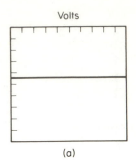

Volts

(a)

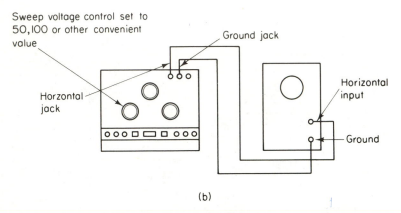

Sweep voltage control set to 50,100 or other convenient value

Ground jack

Horzontal jack

Horizontal input

Ground

(b)

FIGURE 5-6 Alternative method for (a) horizontal calibration with (b) curve tracer.

procedure, but adjust the horizontal gain for two dots per division. That is, the first, third, and fifth dots must coincide exactly, in sequence, with three division markers, as shown in Fig. 5–5.

5-6-6. *High-Voltage Horizontal Calibration*

Adjust the horizontal gain using a high reference voltage as follows.

1. Complete the vertical calibration as described in Sec. 5–6–3.
2. Leave the oscilloscope set for external horizontal input, and the common ground connected between the curve tracer and the oscilloscope, as shown in Fig. 5–6.
3. Disconnect the vertical input from the oscilloscope. Ground the vertical input, if desired.
4. Connect a test lead from the H jack of the curve tracer to the horizontal input of the oscilloscope, as shown in in Fig. 5–6.
5. Set the sweep voltage control to the desired value, such as 50 or 100.

6.   Adjust the horizontal gain of the oscilloscope for a full-scale trace (10 divisions of the graticule). Readjust the horizontal centering as required.

7.   Each graticule division represents 10% of the value at which the sweep voltage control is now set. For example, if the sweep fills all 10 divisions, as shown in Fig. 5–6, and the sweep voltage control is at 50, each division represents 5 V. *Do not* readjust the horizontal gain of the oscilloscope until the horizontal sensitivity is recalibrated to a new value.

5-6-7.   Connecting the Oscilloscope to the Curve Tracer

Figure 5–7 shows the test connections between the oscilloscope and curve tracer during normal operation (after the oscilloscope has been calibrated to the curve tracer). Once the oscilloscope has been calibrated, the calibration steps need not be repeated unless the control settings have been altered. However, an occasional recheck of calibration is suggested.

5-6-8.   Connecting the Semiconductor Device to the Curve Tracer

The curve tracer is designed to accept plug-in transistors, non-plug-in transistors, and diodes. The curve tracer can also be used with a probe for in-circuit tests.

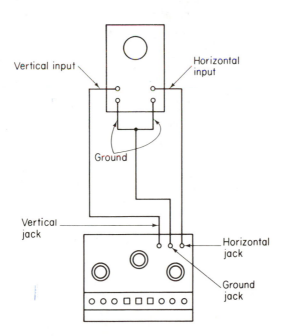

FIGURE 5-7   Connections between the oscilloscope and curve tracer during normal operation.

Plug-In Transistor Connections.   As shown in Fig. 5–8, the curve tracer contains two identical sockets into which most small-signal transistors or FETs may be inserted. Either the left or right socket may be used, as determined by the setting of the socket switch. The socket switch can also be used to start and stop test of the semiconductor (you simply switch to the empty socket to stop the test). This allows complete removal of the test signals while inserting and removing semiconductors from the socket.

The collector, base, and emitter of the transistor (drain, gate, and source of a FET) are inserted into the corresponding pins of the socket (C to collector, D to drain, etc.). For transistors with a grounded case, the fourth pin is inserted into the socket together with the emitter lead, or is externally jumpered to either of the two E (emitter) jacks.

Non-Plug-In Transistor Connections.   As shown in Fig. 5–8, for power transistors, semiconductors with rigid leads, semiconductors with special lead configurations, or for any other reason that prevents the use of plug-in sockets, test leads may be connected from the C-D, B-G, and E-S jacks to the elements of the semiconductor. Two sets of jacks (left and right) are provided. Although either set of jacks may be used, the socket switch activates only the left or right set at a given time (together with the corresponding plug-in socket).

In this particular curve tracer, the jacks, plugs, and probe tips have been color-coded. The color codes are: blue for collector or drain, green for base or gate, and yellow for emitter or source.

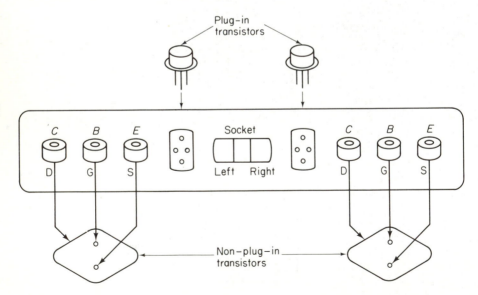

FIGURE 5-8   Connections between semiconductor devices and the curve tracer.

In-Circuit Probe Connections.   Most transistors in consumer and industrial equipment are mounted on printed-circuit (PC) boards with the collector, base, and emitter quite closely spaced. Measurements in such circuits are normally made from *the side opposite the components.* A probe, supplied with the curve tracer, is ideal for in-circuit testing of semiconductors in such circuits. The probe has three tips, which permit contact with the collector, base, and emitter (or drain, gate, and source of a FET) simultaneously, and each one pivots to allow for different spacing. In-circuit transistor testing techniques are discussed further in Chapter 9.

Diode Connections.   As shown in Fig. 5–9, connections to diodes are made using only the collector–emitter (C-E) terminals. Polarity is not of particular importance since polarity is easily reversed using the polarity switch. However, if the cathode of the diode is always connected to the emitter (E) terminal, the polarity switch is set to PNP for reverse bias, and to NPN for forward bias. Reverse bias is applied for testing zener diodes, leakage of signal and rectifier diodes, and inverse peak breakdown voltage, as discussed in Chapter 9. Forward-bias characteristics show voltage drop across the diode junction, as well as diode resistance (or diode open) conditions.

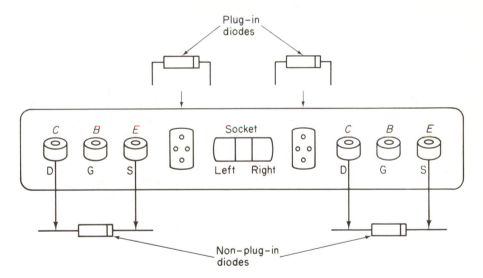

FIGURE 5-9   Connections between diodes and curve tracer.

6

Measuring Voltage and Current

The oscilloscope has both advantages and disadvantages when used to measure voltage and current. The most obvious advantage is that the oscilloscope shows waveform, frequency, and phase simultaneously with the amplitude of the voltage or current being measured. The volt-ohmmeter or electronic voltmeter shows only amplitude. Similarly, most meters are calibrated in relation to sine waves. When the signals being measured contain significant harmonics, the calibrations are inaccurate. With the oscilloscope, the voltage is measured from the displayed wave which includes any harmonic content. In certain applications, the lack of inertia and high-speed response of an oscilloscope make it the only instrument capable of transient voltage measurement.

The only major disadvantage of using an oscilloscope for voltage and current measurement is the problem of resolution. The scales of simple, inexpensive volt-ohmmeters or electronic voltmeters are easier to read than the graticules of an oscilloscope. In most cases, the vertical oscilloscope scales are used for voltage and current measurements, with each scale division representing a given value of voltage or current. Where voltages are large, it is difficult to interpolate between divisions.

Another problem, although not a disadvantage, is that voltages measured with an oscilloscope are peak to peak, whereas most voltages specified in electronic maintenance and troubleshooting manuals are rms. This requires that the peak-to-peak value be converted to rms.

To sum up, if the only value of interest is voltage or current amplitude, use the meter because of its simplicity in readout. Use the oscilloscope where waveshape characteristics are of equal importance to amplitude.

This chapter describes the procedures for measuring both a-c and d-c voltages and currents with an oscilloscope. The screen of any oscilloscope must be calibrated before accurate voltage and current measurements can be made. Laboratory oscilloscopes are calibrated against precision standards. The instruction manuals for laboratory oscilloscopes describe the calibration procedures in great detail. Shop oscilloscopes are often calibrated against any available standard. Often, shop oscilloscope manuals are somewhat sketchy on calibration details. Consequently, this chapter includes calibration procedures directed to shop oscilloscope users, but which can also apply to the laboratory oscilloscope.

NOTE

The vertical amplifier of a laboratory oscilloscope usually has a step attenuator where each step is related to a specific deflection factor (such as V/cm). These oscilloscopes need not be calibrated for voltage or current measurements, since calibration is an internal adjustment performed as part of routine maintenance. The vertical amplifiers of shop oscilloscopes usually have variable attenuators, and possibly a step attenuator. The steps do not, however, have a specific volts/cm deflection factor. Such oscilloscopes must be calibrated before they can be used to measure voltage and current. Therefore, the procedures for both types of oscilloscopes are given in this chapter, where the procedures differ.

6-1. PEAK-TO-PEAK MEASUREMENTS: A-C LABORATORY OSCILLOSCOPE

1. Connect the equipment as shown in Fig. 6–1.
2. Place the oscilloscope in operation (Chapter 5).
3. Set the vertical step attenuator to a deflection factor that will allow the expected signal to be displayed without overdriving the vertical amplifier.
4. Set the input selector to measure ac. Connect the probe to the signal being measured.

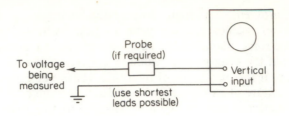

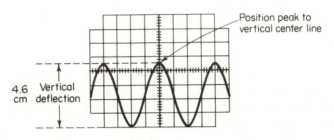

FIGURE 6-1 Measuring peak-to-peak voltages.

5. Switch on the oscilloscope internal recurrent sweep.
6. Adjust the sweep frequency for several cycles on the screen.
7. Adjust the horizontal gain control to spread the pattern over as much of the screen as desired.
8. Adjust the vertical position control so that the downward excursion of the waveform coincides with one of the graticule lines below the graticule centerline, as shown in Fig. 6–1.
9. Adjust the horizontal position control so that one of the upper peaks of the signal lies near the vertical centerline, as shown in Fig. 6–1.
10. Measure the peak-to-peak vertical deflection in centimeters.

NOTE

This technique may also be used to make vertical measurements between two corresponding points on the waveform, other than peak-to-peak. The peak-to-peak points are usually easier to measure.

11. Multiply the distance measured in step 10 by the vertical attenuator switch setting. Also include the attenuation factor of the probe, if any.

Example: Assume a peak-to-peak vertical deflection of 4.6 cm (Fig. 6–1) using a 10× attenuator probe and a vertical deflection factor of 0.5 V/cm.
Use the equation

$$\begin{array}{ccccc} \text{volts} & & \text{vertical} & \text{calibration} & \text{probe} \\ \text{peak to peak} & = & \text{deflection} \times & \text{(V/cm)} \times & \text{attenuation} \\ & & \text{factor} & \text{factor} & \text{factor} \end{array}$$

Substituting the given values yields

$$\text{volts peak to peak} = 4.6 \times 0.5 \times 10 = 23 \text{ V}$$

NOTE

If the voltage being measured is a sine wave, the peak-to-peak value can be converted to peak, rms, or average, as shown in Fig. 6–2. Similarly, if a peak, rms, or average value is given and must be measured on an oscilloscope, Fig. 6–2 can be used to find the corresponding peak-to-peak value.

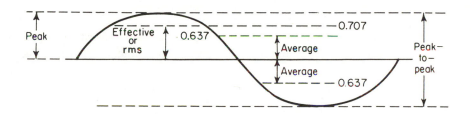

	To Get			
Given	Average	Effective (RMS)	Peak	Peak-to-peak
Average	—	1.11	1.57	1.271
Effective (RMS)	0.900	—	1.411	2.8231
Peak	0.637	0.707	—	2.00
Peak-to-peak	0.3181	0.3541	0.500	—

FIGURE 6-2 Relationship of average, effective, rms, peak, and peak-to-peak values for alternating-current sine waves.

6-2. PEAK-TO-PEAK MEASUREMENTS: A-C SHOP OSCILLOSCOPE

1. Connect the equipment as shown in Fig. 6-1.
2. Place the oscilloscope in operation (Chapter 5).
3. Set the vertical gain control to the calibrate-set position, as determined during the calibration procedure (refer to Sec. 6-5).
4. Set the input selector (if any) to measure ac. Connect the probe to the signal being measured.
5. Switch on the oscilloscope internal recurrent sweep.
6. Adjust the sweep frequency for several cycles on the screen.
7. Adjust the horizontal control to spread the pattern over as much of the screen as desired.
8. Adjust the vertical position control so that the downward excursion of the waveform coincides with one of the graticule lines below the graticule centerline, as shown in Fig. 6-1.

NOTE

Do not move the vertical gain control from the calibrate-set position. Use the vertical position control only.

9. Adjust the horizontal position control so that one of the upper peaks of the signal lies near the vertical centerline, as shown in Fig. 6-1.
10. Measure the peak-to-peak vertical deflection in centimeters.

NOTE

This technique may also be used to make vertical measurements between two corresponding points on the waveform, other than peak to peak. The peak-to-peak points are usually easier to measure.

11. Multiply the distance measured in step 10 by the calibration factor (V/cm) established during calibration. Include the attenuation factor of the probe (if any) and the setting of the attenuator step switch (if any).

Example: Assume a peak-to-peak vertical deflection of 4.6 cm (Fig. 6-1) using a 10× attenuator probe, a 10× position of the step attenuator, and a calibration factor of 0.5 V/cm.

Use the equation

$$\begin{matrix} \text{volts} \\ \text{peak to peak} \end{matrix} = \begin{matrix} \text{vertical} \\ \text{deflection} \\ \text{factor} \end{matrix} \times \begin{matrix} \text{calibration} \\ \text{(V/cm)} \\ \text{factor} \end{matrix} \times \begin{matrix} \text{probe} \\ \text{attenuation} \\ \text{factor} \end{matrix} \times \begin{matrix} \text{step-switch} \\ \text{attenuation} \\ \text{factor} \end{matrix}$$

Substituting the given values yields

$$\text{volts peak to peak} = 4.6 \times 0.5 \times 10 \times 10 = 230 \text{ V}$$

NOTE

If the voltage being measured is a sine wave, the peak-to-peak value can be converted to peak, rms, or average, as shown in Fig. 6–2. If a peak, rms, or average value is given and must be measured on an oscilloscope, Fig. 6–2 can be used to find the corresponding peak-to-peak value.

6-3. *INSTANTANEOUS VOLTAGE MEASUREMENTS: D-C LABORATORY OSCILLOSCOPE*

1. Connect the equipment as shown in Fig. 6–3.
2. Place the oscilloscope in operation (Chapter 5).
3. Set the vertical step attenuator to a deflection factor that will allow the expected signal, plus any dc, to be displayed without overdriving the vertical amplifier.
4. Set the input selector to ground.

NOTE

On most laboratory oscilloscopes, the input switch that selects either a-c or d-c measurement also has a position that connects both vertical input terminals to ground (or shorts them together). If no such switch position is provided, short the vertical input terminals by connecting the probe (or other lead) to ground.

5. Switch on the oscilloscope internal recurrent sweep. Adjust the horizontal gain control to spread the trace over as much of the screen as desired.
6. Using the vertical position control, position the trace to a line of the graticule below the centerline, as shown in Fig. 6–3. This establishes the reference line. If the average signal (a-c plus d-c) is

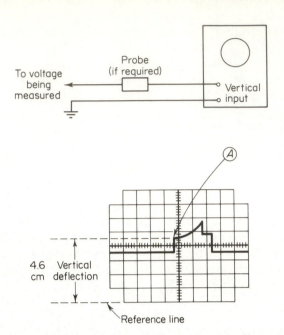

FIGURE 6-3 Measuring instantaneous (or d-c) voltages.

negative with respect to ground, position the trace to a reference line above the graticule center line. *Do not* move the vertical position control after this reference line has been established.

NOTE

To measure a voltage level with respect to a voltage other than ground, make the following changes in steps 4 and 6: set the input selector to measure dc; apply the reference voltage to the vertical input; then position the trace to the reference line.

7. Set the input selector switch to dc. The ground reference line, if used, can be checked at any time by switching the input selector to the ground position. Connect the probe to the signal being measured.

8. If the waveform is outside the viewing area, set the vertical step attenuator so that the waveform is visible.

9. Adjust the sweep frequency and horizontal gain controls to display the desired waveform.

10. Measure the distance in centimeters between the reference line and the point on the waveform at which the d-c level is to be measured.

For example, in Fig. 6-3, the measurement is made between the reference line and point *A*.

11. Establish polarity of the signal. Any signal-inverting switches on the oscilloscope must be in the normal position. If the waveform is above the reference line, the voltage is positive; below the line, negative.

12. Multiply the distance measured in step 10 by the vertical attenuator switch setting. Also include the attenuation factor of the probe, if any.

Example: Assume that the vertical distance measured is 4.6 cm (Fig. 6-3). The waveform is above the reference line, using a 10× attenuator probe and a vertical deflection factor of 2 V/cm.

Use the equation

$$\begin{array}{c} \text{instantaneous} \\ \text{voltage} \end{array} = \begin{array}{c} \text{vertical} \\ \text{distance} \\ \text{(cm)} \end{array} \times \text{polarity} \times \begin{array}{c} \text{calibration} \\ \text{(V/cm)} \\ \text{factor} \end{array} \times \begin{array}{c} \text{probe} \\ \text{attenuation} \\ \text{factor} \end{array}$$

Substituting the given values yields

$$\text{instantaneous voltage} = 4.6 \times +1 \times 2 \times 10$$

The instantaneous voltage is +92 V.

6-4. *INSTANTANEOUS VOLTAGE MEASUREMENTS: D-C SHOP OSCILLOSCOPE*

1. Connect the equipment as shown in Fig. 6-3.

2. Place the oscilloscope in operation (Chapter 5).

3. Set the vertical gain control to the calibrate-set position, as determined during the calibration procedure (refer to Sec. 6-5).

4. Set the input selector (if any) to ground. If no such switch position is provided, short the vertical input terminals by connecting the probe (or other lead) to ground.

5. Switch on the oscilloscope internal recurrent sweep. Adjust the horizontal gain control to spread the trace over as much of the screen as desired.

6. Using the vertical position control, position the trace to a line of the graticule below the graticule centerline, as shown in Fig. 6-3. This establishes the reference line. If the average signal (a-c plus d-c) is negative with respect to ground, position the trace to a

reference line above the graticule centerline. *Do not* move the vertical position control after this reference line has been established.

NOTE

To measure a voltage level with respect to a voltage other than ground, make the following changes in steps 4 and 6: instead of shorting the vertical input terminals, apply the reference voltage to the vertical input; then position the trace to the reference line.

7. Set the input selector (if any) to dc. (The oscilloscope must be capable of d-c measurement. Some shop-type oscilloscopes do not have this capability.) The ground reference line, if used, can be checked at any time by switching the input selector to the ground position. Connect the probe to the signal being measured.

8. If the waveform is outside the viewing area, set the vertical step attenuator (if any) so that the waveform is visible. Do not move the vertical position or vertical gain controls to bring the waveform into view.

9. Adjust the sweep frequency and horizontal gain controls to display the desired waveforms.

10. Measure the distance in centimeters between the reference line and the point on the waveform at which the d-c level is to be measured. For example, in Fig. 6-3, the measurement is made between the reference line and point *A*.

11. Establish polarity of the signal. Any signal-inverting switches on the oscilloscope must be in the normal position. If the waveform is above the reference line, the voltage is positive; below the line, negative.

12. Multiply the distance measured in step 10 by the calibration factor (V/cm) established during calibration. Also include the attenuation factor of the probe, if any, and the setting of the attenuator step switch, if any.

Example: Assume that the vertical distance measured is 4.6 cm (Fig. 6-3). The waveform is above the reference line, using a 20× attenuator probe, a 10× position of the step attenuator, and a calibration factor of 2 V/cm. Use the equation

instantaneous voltage

= vertical distance (cm) × polarity × calibration (V/cm) factor × probe attenuation factor × step-switch attenuation factor

Substituting the given values yields

$$\text{instantaneous voltage} = 4.6 \times + 1 \times 2 \times 10 \times 10$$

The instantaneous voltage is $+920$ V.

6–5. CALIBRATING THE VERTICAL AMPLIFIER FOR VOLTAGE MEASUREMENTS

On those laboratory oscilloscopes that have a vertical step attenuator related to some specific deflection factor (V/cm), the calibration procedure is an internal adjustment accomplished as part of routine maintenance. On other oscilloscopes, the vertical amplifier must be calibrated for voltage measurements. The basic calibration procedure consists of applying a reference voltage of known amplitude to the vertical input and adjusting the vertical gain control for specific deflection. Then the reference voltage is removed and the test voltages are measured, *without* changing the vertical gain setting. The calibration will remain accurate as long as the vertical gain control is at this calibrate-set position.

The vertical amplifier can be voltage-calibrated by several methods. For example, the calibrating voltage can be a-c or d-c, variable or fixed, internal or external. The method used depends upon the type of oscilloscope and the available calibrating voltage. The following sections describe each of the methods in turn.

6–5–1. Voltage Calibration with External Dc

On oscilloscopes that do not have internal voltage reference sources, it is necessary to use an external calibrator. Any d-c source of known accuracy can be used. It is best to select an approximate calibrating voltage value that will produce at least half-scale deflection with the vertical gain control near midscale and the step attenuator (if any) is the $\times 1$ position. (This permits the step-attenuator multiplier function to be used.)

NOTE

The important point to remember concerning the external calibrating voltage is its accuracy. Accuracy of the oscilloscope voltage measurements will be no greater than the accuracy of the calibrating voltage.

1.   Connect the equipment as shown in Fig. 6–4.
2.   Place the oscilloscope in operation (Chapter 5).

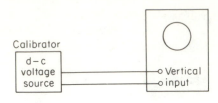

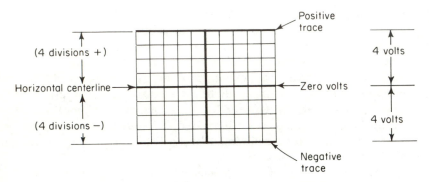

FIGURE 6-4 Voltage calibration with external dc.

3. Using the vertical position control, position the trace to the graticule horizontal centerline.

NOTE

Switch the internal recurrent sweep on (for a line trace), or off (for a dot), whichever trace is most convenient for calibration.

4. Turn on the calibrator and set the calibrating voltage to the desired calibrating value.

NOTE

The exact value of the calibration voltage depends upon the graticule scale. For example, if there are eight vertical divisions (four above and four below the horizontal centerline) a value of 0.4, 4, or 40 V would be convenient.

5. Without touching the vertical position control, set the vertical gain control to move the dot or line trace vertically up to the desired number of screen divisions. For example, assuming a calibrating

voltage of 4 V and a scale as shown in Fig. 6–4, set the vertical gain control so that the trace is moved up to the top line, or four divisions from the centerline. Thus, each division will equal 1 V. In the example shown in Fig. 6–4, this would give the oscilloscope a vertical deflection factor of 1 V/cm, with the step attenuator set to × 1. If the step attenuator were then moved to × 10, the factor would be 10 V/cm.

NOTE

If the external source is fixed, the process must be reversed, and a scale factor must be selected to match the voltage. For example, assume that the only calibrating source is a 1.5-V battery of known accuracy. The vertical gain control could then be set to provide a deflection of three divisions (from the horizontal center line up three divisions). This would give a vertical deflection factor of 0.5 V/cm, with the step attenuator set to × 1. If the attenuator were moved to × 10, the factor would be 5 V/cm.

6. Remove the calibrating voltage and check that the trace is returned to the horizontal centerline.
7. Reverse the calibrating voltage leads. Reapply the voltage and check that the trace is moved *below* the horizontal centerline by the same number of divisions as obtained in step 5.

NOTE

Thus far, the procedures have provided calibration for measurement of both positive and negative d-c voltages (positive voltages are measured above the horizontal center line; negative voltages below the centerline). Where the voltages to be measured are known to be all positive (or negative), it may prove convenient to use the entire graticule scale. In that event, use the vertical position control to position the trace to the bottom graticule horizontal line (for all positive voltages), or the top horizontal line (for all negative voltages). Then apply the calibrating voltage and set the vertical gain control to move the trace vertically up (or down) the desired number of screen divisions.

8. If the calibrating source voltage is variable, check the accuracy of the calibration by applying various voltages, at various settings of the step attenuator (if any).

NOTE

If the d-c calibrating voltage is not variable, it may be made so, using the circuit of Fig. 6–5. The accuracy of such a test configuration is entirely dependent upon the accuracy of meter M_1.

9. If desired, the position of the vertical gain control should be noted and recorded as the calibrate-set position. Use this same position for all future voltage measurements. It is recommended that the calibration be checked at frequent intervals.

NOTE

Once the calibrate-set position has been established, the reference line (calibration voltage removed, zero volts) can be moved up or down as required by the vertical position control, without affecting the V/cm factor. Remember, however, that any voltage measurements must be made from the reference line. For example, assume that the horizontal centerline is used as the reference line during calibration, and the deflection factor is established as 1 V/cm. Then, during actual voltage measurement, the reference line is moved down 2 cm below the centerline, and a voltage is measured 3 cm above the centerline. Since the voltage measured is 5 cm from the reference line, the correct reading would be 5 V.

6–5–2. Voltage Calibration with External AC

On those oscilloscopes that do not have internal voltage reference sources it is necessary to use an external calibrator. Any a-c source of known accuracy can be used. It is best to select an approximate calibrating voltage value that will produce near full-scale deflection with the vertical gain control near midscale, and the step attenuator (if any) in the ×1 position. (This will permit the step-attenuator multiplier function to be used.)

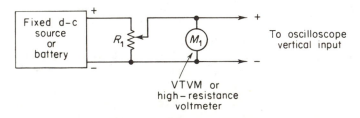

FIGURE 6-5   Circuit for d-c calibration.

NOTE

The important point to remember concerning the external calibrating voltage is its accuracy. Accuracy of the oscilloscope voltage measurements will be no greater than the accuracy of the calibrating voltage. Another point to remember is that the oscilloscope display is usually calibrated for peak-to-peak voltage, whereas the meter or other device indicating the calibrating voltage will probably be in rms. If the calibrating voltage is a sine wave, the rms value can be converted to peak to peak, as shown in Fig. 6–2. If the external calibrating voltage is a square wave or pulse, its value will be peak to peak.

1. Connect the equipment as shown in Fig. 6–6.
2. Place the oscilloscope in operation (Chapter 5).
3. Using the vertical position control, position the trace to the graticule horizontal centerline.

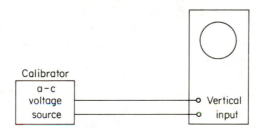

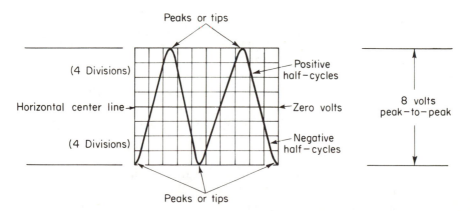

FIGURE 6-6 Voltage calibration with external ac.

NOTE

Switch the internal recurrent sweep on (for a normal line trace), or off (for a dot), whichever trace is most convenient for calibration. If the internal sweep is not on, the trace will appear as a vertical line when the calibrating voltage is applied.

4. Turn on the calibrator and set the calibrating voltage to the desired calibrating value.

NOTE

The exact value of the calibrating voltage depends upon the graticule scale. For example, if there are eight vertical divisions (four above and four below the horizontal centerline), a value of 0.8, 8, or 80 V would be convenient.

5. Without touching the vertical position control, adjust the vertical gain control to align tips of positive half-cycles and tips of negative half-cycles with the desired scale divisions. For example, assuming a calibrating voltage of 8 V peak to peak, and a scale as shown in Fig. 6–6, set the vertical gain control so that the trace is spread from the top line (four divisions up from the centerline) to the bottom line (four divisions below the centerline). Thus, each division will equal 1 V peak to peak. In the example shown in Fig. 6–6, this would give the oscilloscope a vertical deflection factor of 1 V/cm, with the step attenuator set to ×1. If the step attenuator were then moved to ×10, the factor would be 10 V/cm.

NOTE

If the external source is fixed, the process must be reversed, and a scale factor must be selected to match the voltage. For example, assume that the only calibrating source is a 1-V ac (rms) of known accuracy. This is equal to 2.828 V peak to peak. The vertical gain control could then be set to provide a spread of slightly less than three divisions (2.828 divisions). This would give a vertical deflection factor of 1 V/cm peak to peak, with the step attenuator set to ×1. If the attenuator were moved to ×10, the factor would be 10 V/cm.

6. If the calibrating source voltage is variable, check the accuracy of the calibration by applying various voltages, at various settings of the step attenuator (if any).

NOTE

If the a-c calibrating voltage is not variable, it may be made so, using the circuit of Fig. 6–7. The accuracy of such a test configuration is entirely dependent upon the accuracy of meter. M_1.

7. If desired, the position of the vertical gain control should be noted and recorded as the calibrate-set position. Use this same position for all future voltage measurements. It is recommended that the calibration be checked at frequent intervals.

NOTE

Once the calibrate-set position has been established the trace can be moved up or down as required by the vertical position control, without affecting the V/cm factor.

6–5–3. *Voltage Calibration with Internal Ac*

Most oscilloscopes have an internal voltage source of known amplitude and accuracy available for calibration. On some oscilloscopes, this calibrating voltage is available from terminals or a jack on the front panel. On other oscilloscopes, the calibrating voltage is applied to the vertical input when one of the controls (usually the vertical input selector) is set to the calibrate position.

1. Connect the equipment as shown in Fig. 6–8, if the calibrating voltage is available at the front panel. If the calibrating voltage is applied by means of a control, set the control to the calibrate position.
2. Place the oscilloscope in operation (Chapter 5).

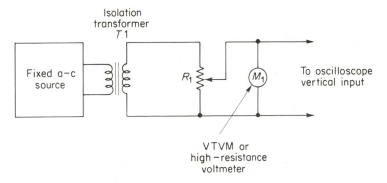

FIGURE 6-7   Circuit for a-c calibration.

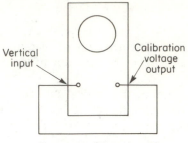

Temporary interconnection

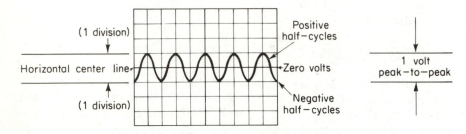

FIGURE 6-8 Voltage calibration with internal ac.

3. With the calibrate voltage removed (front-panel connections temporarily removed, or vertical input control set to normal), use the vertical position control and position the trace to the graticule horizontal centerline.

NOTE

Switch the internal recurrent sweep on (for a normal line trace), or off (for a dot), whichever trace is most convenient for calibration. If the internal sweep is not on, the trace will appear as a vertical line when the calibrating voltage is applied.

4. Apply the calibrating voltage.
5. Without touching the vertical position control, adjust the vertical gain control to align tips of positive half-cycles and tips of negative half-cycles with the desired scale divisions. For example, assume a calibrating voltage of 1 V peak to peak (which is typical for many oscilloscopes), set the vertical gain control so that the trace is spread from one division above the centerline to one division below the

centerline. Thus, each division will equal 0.5 V peak to peak. In the example shown in Fig. 6–8, this would give the oscilloscope a vertical deflection factor of 0.5 V/cm, with the step attenuator set to × 1. If the step attenuator were then moved to × 10, the factor would be 5 V/cm.

6.   If desired, the position of the vertical gain control should be noted and recorded as the calibrate-set position. Use this same position for all future voltage measurements. It is recommended that the calibration be checked at frequent intervals.

NOTE

Once the calibrate-set position has been established the trace can be moved up or down as required by the vertical position control, without affecting the V/cm factor.

6–5–4. Voltage Calibration with Internal Square Waves

Some oscilloscopes have an internal square-wave source available for calibration. Usually, this source is variable in amplitude and is adjusted by a front-panel control. The square-wave amplitude is read off a scale on the amplitude adjustment control. The square waves are applied to the vertical input when one of the controls (usually the vertical input selector) is set to calibrate position.

1.   Place the oscilloscope in operation (Chapter 5).
2.   With the calibrate square waves removed (vertical input control set to normal), use the vertical position control and position the trace to the graticule horizontal centerline.
3.   Apply the calibrating square waves.
4.   Set the horizontal sweep frequency to some frequency lower than that of the internal calibrating square waves.

NOTE

Several cycles of square waves should appear when the sweep frequency is lower than the internal square-wave calibrating frequency. For example, four square waves should appear if the sweep frequency is one-fourth of the calibrating frequency. One square wave should appear if the sweep frequency and calibrating frequency are the same. If the sweep frequency is considerably lower than the calibrating frequency, the flat peaks of the square waves will blend and appear as two horizontal lines. If the internal sweep is not on, the trace will appear as a vertical line when the calibrating square waves are applied.

5. Without touching the vertical position control, adjust vertical gain control and align the flat peaks of the square waves with the desired scale divisions. For example, assume square waves with a peak-to-peak amplitude of 6 V and a scale, as shown in Fig. 6–9. Set the vertical gain control so that the square waves are spread a total of six divisions (three divisions up from the centerline and three divisions below the centerline). Thus, each division will equal 1 V peak to peak. In the example shown in Fig. 6–9, this would give the oscilloscope a vertical deflection factor of 1 V/cm, with the step attenuator set to × 1. If the step attenuator were then moved to × 10, the factor would be 10 V/cm.

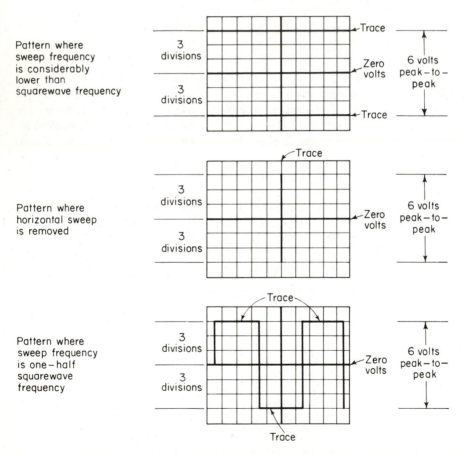

FIGURE 6-9 Voltage calibration with internal square waves.

NOTE

When the graticule horizontal centerline is used as a reference, as described in step 2 of this procedure, the square-wave peaks above the centerline indicate positive voltage, whereas peaks below the centerline indicate negative voltage. The amplitude of the square waves above (and below) the centerline is equal to peak a-c voltage. This is equivalent to one-half the peak-to-peak voltage (Fig. 6–2).

6. If desired, the position of the vertical gain control should be noted and recorded as the calibrate-set position. Use this same position for all future voltage measurements. It is recommended that the calibration be checked at frequent intervals.

NOTE

Once the calibrate-set position has been established, the trace can be moved up or down as required by the vertical position control without affecting the V/cm factor.

6–6. VOLTAGE MEASUREMENTS WITH A VARIABLE CALIBRATING SOURCE

When an oscilloscope has an internal variable calibrating voltage source, or an accurate external variable source is readily available, it is often convenient to measure voltages using the variable calibrator. This method is sometimes known as *indirect voltage measurement*. Its primary advantage is that the oscilloscope vertical amplifier and graticule screen need not be precalibrated, and the vertical gain control can be set to any convenient level. The basic procedure consists of measuring the test voltage and noting the number of divisions occupied by the trace. Then the test voltage is removed and replaced by the calibrating voltage. The calibrating voltage is adjusted until it occupies the same number of divisions as the test voltage. The calibrating voltage amplitude is read off the corresponding oscilloscope control, or the meter of the external calibrator. This method is quite accurate, but it does require continued switching back and forth between test and calibrating voltages.

The method has some other disadvantages. If the calibrating voltage is internal, it may be fed directly to the vertical amplifier input when the appropriate control is set to a calibrate position, as described in Sec. 6–5–4. On such oscilloscopes, when a probe is used, it is necessary to include the probe multiplication factor. For example, assume that a voltage is measured through a 10:1 probe, that the test signal occupies three divisions, that the oscilloscope's internal square-wave calibrating signal is switched on and adjusted to occupy the same three divisions, and that the calibration signal-amplitude control indicates 7 V. Since the internal calibration voltage is applied directly to the vertical input, and the test voltage is applied through a 10:1 probe, the actual test voltage must be 70 V.

Another major disadvantage is that the calibrating voltage source, internal or external, may not be equal to the test voltage. This condition can be offset by means of the vertical step attenuator, or by a voltage-divider probe. For example, if the test signal may reach 300 V, whereas the calibration source has a maximum of 30 V, use a 10:1 voltage-divider probe, or the × 10 position of the vertical step attenuator. Measure the test voltage with the probe (or the step attenuator in × 10); then apply the calibrating voltage with the probe removed (or the step attenuator in × 1) and adjust for the same number of divisions. Make certain to multiply the calibration voltage control indication by the appropriate factor.

NOTE

In the following procedure, it is assumed that the oscilloscope has an internal variable square-wave calibrating voltage and that the voltage to be measured is a sine wave.

1. Connect the equipment as shown in Fig. 6–10.
2. Place the oscilloscope in operation (Chapter 5).
3. Set the vertical step attenuator and/or vertical gain control to a deflection factor that will allow the expected signal to be displayed without overdriving the vertical amplifier.
4. Set the input selector (if any) to measure ac. Connect the probe to the signal being measured.
5. Switch on the oscilloscope internal recurrent sweep.
6. Adjust the sweep frequency for several cycles on the screen.
7. Adjust the horizontal gain control to spread the pattern over as much of the screen as desired.
8. Adjust the vertical position control so that the downward excursion of the waveform coincides with one of the graticule lines below the graticule centerline, as shown in Fig. 6–10. Adjust the vertical gain control so that the pattern occupies an exact, easily measured number of divisions.
9. Adjust the horizontal position control so that one of the upper peaks of the signal lies near the vertical centerline, as shown in Fig. 6–10.
10. Without disturbing any of the controls, switch on the internal calibrating square-wave signals. The test signal should be removed from the screen and replaced by the square-wave calibrating signals.

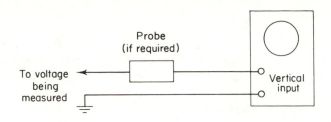

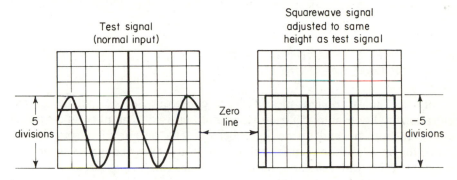

FIGURE 6-10 Indirect voltage measurement with variable square-wave calibrating signal.

11. Adjust the square-wave calibrating signal amplitude until the square-wave pattern occupies the same number of divisions as the test voltage.

NOTE

If the square-wave signal is insufficient and cannot be adjusted to occupy the same number of divisions, switch the vertical step attenuator to a convenient multiplier scale. *Do not* move the vertical gain control.

12. Read the voltage off the calibrator amplitude control. Multiply the voltage by any probe and/or step-attenuator attenuation factor.

Example: Assume a peak-to-peak deflection factor, as shown in Fig. 6–10, using a 10× attenuator probe. Assume, too, that the calibrator amplitude control reads 3.7 V. Further assume that the test voltage was measured with the step attenuator in ×10, and the square waves were adjusted to the same height with the step attenuator in ×1.

Use the equation

$$\text{volts peak to peak} = \begin{array}{c}\text{calibrator}\\\text{control}\\\text{reading}\end{array} \times \begin{array}{c}\text{probe}\\\text{attenuation}\\\text{factor}\end{array} \times \begin{array}{c}\text{difference in}\\\text{step-attenuator}\\\text{factor}\end{array}$$

Substituting the given values yields

$$\text{volts peak to peak} = 3.7 \times 10 \times 10 = 370 \text{ V}$$

6-7. COMPOSITE AND PULSATING VOLTAGE MEASUREMENTS

In practice, most voltages measured are composites of ac and dc, or are pulsating dc. For example, a transistor amplifier used to amplify an a-c signal will have both ac and dc on its collector. The output of a rotating d-c generator or solid-state rectifier will be pulsating, even though its polarity is constant. Such composite and pulsating voltages can be measured quite readily on an oscilloscope capable of measuring dc (as are most laboratory oscilloscopes). The procedures are essentially a combination of peak-to-peak measurements (Secs. 6–1 and 6–2) and instantaneous d-c measurements (Secs. 6–3 and 6–4). Composite and pulsating voltages can also be measured by the indirect method (Sec. 6–6). This is usually difficult. Also, since most oscilloscopes capable of measuring dc are also direct reading, the indirect method will not be discussed.

1. Connect the equipment as shown in Fig. 6–11.
2. Place the oscilloscope in operation (Chapter 5).
3. Set the vertical step attenuator to a deflection factor that will allow the expected signal, plus any dc, to be displayed without overdriving the vertical amplifier.
4. Set the input selector to ground.

NOTE

On most laboratory oscilloscopes, the input switch that selects either a-c or d-c measurement also has a position that connects both vertical input terminals to ground (or shorts them together). If such switch position is not provided, short the vertical input terminals by connecting the probe (or other lead) to ground.

5. Switch on the oscilloscope internal recurrent sweep. Adjust the

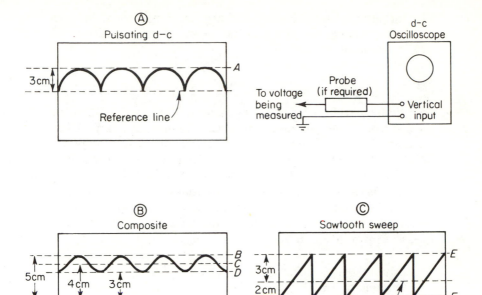

FIGURE 6-11 Measurement of composite and pulsating voltages.

horizontal gain control to spread the trace over as much of the screen as desired.

6. Using the vertical position control, position the trace to a convenient location on the graticule screen. If the voltage to be measured is pulsating dc, the horizontal centerline should be convenient. If the voltage is a composite, and the average signal (ac plus dc) is positive, position the trace below the centerline. If the average is negative, position the trace above the centerline. *Do not move the vertical position control after this reference line has been established.*

7. Set the input selector switch to dc. The ground reference line can be checked at any time by switching the input selector to the ground position. Connect the probe to the signal being measured.

8. If the waveform is outside the viewing area, set the vertical step attenuator so that the waveform is visible.

9. Adjust the sweep frequency and horizontal gain controls to display the desired waveform.

10. Establish polarity of the signal. Any signal-inverting switches on the oscilloscope must be in the normal position. If the waveform is above the reference line, the voltage is positive; below the line,

negative. Measure the distance in centimeters between the reference line and the point on the waveform at which the level is to be measured.

NOTE

If the voltage to be measured is pulsating dc, the trace will remain on one side of the reference line, but will start and stop at the reference line, as shown in Fig. 6–11a. If the voltage is a composite of ac and dc, the trace may be on either side of the reference line; it may possibly cross over the reference line, but usually remains on one side, as shown in Fig. 6–11b. If the voltage is a non-sine wave (such as sawtooth, pulse, spike, etc.), the trace may appear on both sides of the reference line, or may be displaced above or below the line, as shown in Fig. 6–11c.

11.   Multiply the distance measured in step 10 by the vertical attenuator switch setting. Also include the attenuation factor of the probe, if any.

Example: Assume that the vertical distance measured is 3 cm from the reference line to point A of Fig. 6–11a. The waveform is above the reference line (pulsating dc), using a $10\times$ attenuator probe and a vertical deflection factor of 2 V/cm.
Substituting the given values yields

peak of the pulsating d-c voltage $= 3 \times + 1 \times 2 \times 10 = 60$ V (peak)

Example: Assume that the vertical distance measured is 3 cm from the reference line to point D (Fig. 6–11b), 4 cm to point C (Fig. 6–11b), and 5 cm to point B (Fig. 6–11b). The waveform is above the reference line (ac combined with positive dc), using a $10\times$ attenuator probe and a vertical deflection factor of 2 V/cm.
Substituting the given values yields

d-c component (reference line to point C) $= 4 \times + 1 \times 2 \times 10$
$$= +80 \text{ V}$$
peak to peak of a-c component (point B to D)
$$= 2 \times 2 \times 10 = 40 \text{ V (peak to peak)}$$

NOTE

The 2-cm value is obtained by subtracting the point D value (3 cm) from the point B value (5 cm).

Example: Assume that the vertical distance measured is 3 cm from the reference line to point E (Fig. 6–11c) and 2 cm from the reference line to point F. The waveform is above and below the reference line (sawtooth sweep), using a $10\times$ attenuator probe and a vertical deflection factor of 2 V/cm.

Substituting the given values,

positive peak of sweep (point E)

$$= 3 \times +1 \times 2 \times 10 = +60 \text{ V (peak)}$$

negative peak of sweep (point F)

$$= 2 \times -1 \times 2 \times 10 = -40 \text{ V (peak)}$$

peak to peak of sweep (point E to F)

$$= 60 + 40 = 100 \text{ V (peak to peak)}$$

6–8. VOLTAGE COMPARISON MEASUREMENTS: LABORATORY OSCILLOSCOPES

In some applications, it may be necessary to establish a set of deflection factors other than those indicated by the vertical step-attenuator switch. This is useful for comparing signals to a reference voltage amplitude. To establish a new set of deflection factors based upon a specific reference amplitude, the vertical amplifier must have a variable gain control, as well as the step attenuator.

1. Connect the equipment as shown in Fig. 6–1.
2. Place the oscilloscope in operation (Chapter 5).
3. Set the vertical step attenuator to a deflection factor that will allow the expected signal to be displayed without overdriving the vertical amplifier.
4. Apply the reference signal of known value to the vertical input.
5. Switch on the oscilloscope internal recurrent sweep. Adjust the sweep frequency for several cycles on the screen. Adjust the horizontal gain control to spread the pattern over as much of the screen as desired.
6. Using both the vertical step attenuator and variable gain control, adjust the display for an exact number of vertical centimeter divisions. *Do not* move the variable gain control after obtaining the desired deflection.
7. Divide the amplitude of the reference signal (volts) by the product of the deflection in centimeters (established in step 6) and the ver-

tical attenuator switch setting. This is the *deflection conversion factor*.

$$\begin{matrix} \text{deflection} \\ \text{conversion} \\ \text{factor} \end{matrix} = \frac{\text{reference signal amplitude (V)}}{\text{deflection (cm)} \times \text{attenuator switch setting}}$$

8. To establish an adjusted deflection factor at any setting of the vertical attenuator switch, multiply the attenuator switch setting by the deflection conversion factor established in step 7.

$$\begin{matrix} \text{adjusted} \\ \text{deflection} \\ \text{factor} \end{matrix} = \begin{matrix} \text{attenuator} \\ \text{switch} \\ \text{setting} \end{matrix} \times \begin{matrix} \text{deflection} \\ \text{conversion} \\ \text{factor} \end{matrix}$$

This adjusted deflection factor applies only if the variable vertical gain control is not moved from the position established in step 6.

9. To determine the peak-to-peak amplitude of a signal compared to a reference, disconnect the reference and set the vertical step attenuator to a deflection factor that will provide sufficient deflection to make the measurement. *Do not* move the variable gain control from the position set in step 6.

10. Apply the signal to the vertical input and measure the deflection. The amplitude may be determined by the following equation:

$$\begin{matrix} \text{signal} \\ \text{amplitude} \end{matrix} = \begin{matrix} \text{adjusted} \\ \text{deflection} \\ \text{factor} \end{matrix} \times \text{deflection (cm)}$$

Example: Assume a reference signal amplitude of 30 V, a vertical attenuator switch setting of 5, and a deflection of 4 cm. Substituting these values in the deflection conversion factor formula (step 7) yields

$$\begin{matrix} \text{deflection} \\ \text{conversion} \\ \text{factor} \end{matrix} = \frac{30}{4 \times 5} = 1.5$$

Then, with a step-attenuator setting of 10, the adjusted deflection factor (step 8) would be

$$\begin{matrix} \text{adjusted} \\ \text{deflection} \\ \text{factor} \end{matrix} = 10 \times 1.5 = 15 \text{ V/cm}$$

To determine the peak-to-peak amplitude of an applied signal that produces a vertical deflection of 5 cm, use the signal amplitude formula (step 10):

$$\text{signal amplitude} = 15 \times 5 = 75 \text{ V}$$

6–9. CURRENT MEASUREMENTS WITH A TEST RESISTOR

The most common method of measuring an unknown current is passing it through a resistance of known value, and then measuring the resultant voltage. This is the basic principle of most voltmeters. Since the oscilloscope can be used as a voltmeter, it can also be adapted to measure current. A resistor of known value and accuracy is the only other component required for the procedure. Once the voltage has been measured, the current can be calculated using the basic Ohm's law equation $I = E/R$.

1. Connect the equipment as shown in Fig. 6–12.
2. Place the oscilloscope in operation (Chapter 5).
3. Apply the current through the resistor.

NOTE

If a 1-Ω resistor is used, the calculations will be simplified. The unknown current will be equal to the measured voltage. The wattage of the resistor must be at least double the square of the

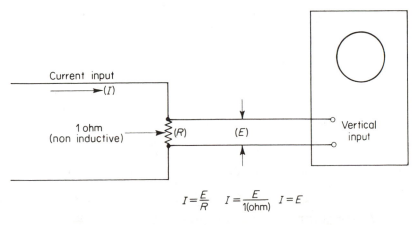

$$I = \frac{E}{R} \qquad I = \frac{E}{1(\text{ohm})} \qquad I = E$$

FIGURE 6–12 Current measurements with a test resistor.

maximum current (in amperes). For example, if the maximum anticipated current is 10 A, the minimum wattage of the resistor should be $10^2 \times 2 = 200$ W.

4.   Measure the voltage drop across the resistor using the procedures of Secs. 6–1 through 6–4, 6–6, and 6–7, whichever is applicable to the type of voltage being measured. The unknown current will be equal to the voltage measured, provided that a 1-Ω resistance is used.

NOTE

If peak-to-peak voltage is measured, the resultant current value will also be peak to peak. To determine the rms or average value, convert the measured voltage to rms or average using Fig. 6–2.

Example: Assume that the voltage drop across the 1-Ω test resistor is 10 V peak to peak.

$$10 \text{ V peak to peak} = 3.535 \text{ V rms}$$

$$I = \frac{3.535}{1} = \text{rms current } 3.535 \text{ A}$$

5.   If the current being measured is the result of a composite voltage (ac plus dc), both the a-c and d-c voltages should be measured separately, as described in Sec. 6–7. The a-c voltage (peak to peak) should then be converted to rms (using Fig. 6–2). The d-c and a-c currents (which are equivalent to the corresponding voltages) should be combined to find the composite or total current as follows:

$$\text{total current} = \sqrt{\text{a-c current (rms)}^2 + \text{d-c current}^2}$$

6-10.   CURRENT MEASUREMENTS WITH A CURRENT PROBE

Measuring current with a test resistor as described in Sec. 6–9 has two disadvantages: (1) the circuit must be opened so that the resistor can be inserted during the test, and (2) operation of the circuit can be affected by the additional resistance. Both of these problems can be eliminated by means of a current probe. Such probes can be obtained as accessories for most laboratory oscilloscopes.

Current probes operate on the same basic principle as the clamp-type ammeters used in power electrical equipment. The basic element of a current probe is the ferrite core, which is attached to a handle. Ferrite material is used to provide a wide frequency response. The core, shown functionally in Fig.

6–13, is designed to be opened and closed so that it may be clamped around the wire carrying the current to be measured. The wire forms the primary of a transformer. The secondary is formed by the probe pickup coil. Current passing through the wire induces a voltage in the secondary. The secondary output voltage is usually quite high in respect to the primary, since the pickup coil has many turns.

The probe output voltage is applied to the oscilloscope vertical input and is measured in the normal manner. Since there is a direct relationship between voltage and current, the current can be calculated from the voltage indicated on the oscilloscope. If the vertical amplifier is calibrated for a given value of reference deflection, the current may be read directly from the oscilloscope screen. For example, if the probe output is 1 mV per mA of current, and the oscilloscope vertical amplifier is calibrated from 1 mV/cm, the current may be read directly from the scale (1mA/cm).

Very often, current probes are used with an amplifier, since the probe output is quite low in relation to the average oscilloscope deflection factor. A typical probe has an output of 1 mV/mA, whereas the average laboratory oscilloscope may have a vertical sensitivity of 5 mV/cm. Thus, it would require 50 mA to produce a deflection of 1 cm.

Since current probes are used with laboratory oscilloscopes and are provided with detailed instructions for their use, no operating procedures are given here.

6–11. DIFFERENTIAL MEASUREMENTS WITH A DUAL-TRACE OSCILLOSCOPE

A dual-trace oscilloscope with an ADD function (both channels added algebraically) can be used to make differential measurements. Such oscilloscopes may be used to observe waveforms and measure voltages be-

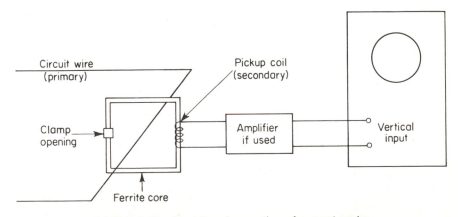

FIGURE 6-13 Functional operating of current probe.

tween two points in a circuit, neither of which is at circuit ground. Figure 6–14 shows the connections for a typical differential measurement (to measure the output signal of a push–pull amplifier). Other differential measurements include monitoring the inputs or outputs of a differential amplifier, the output of a phase splitter, and the amount of signal developed across a single section of a voltage divider or attenuator.

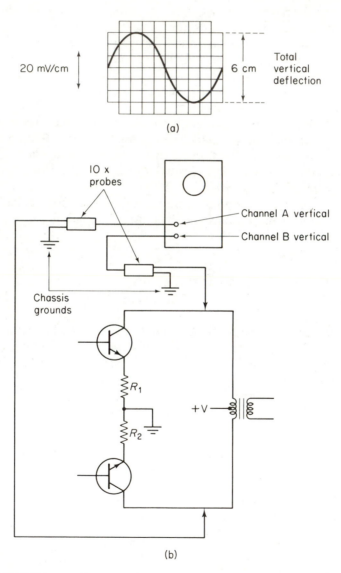

FIGURE 6-14 Measuring differential voltages with a dual-trace oscilloscope.

1. Connect the equipment as shown in Fig. 6–14.
2. Place the oscilloscope in operation (Chapter 5).
3. Connect the ground clips of the two oscilloscope probes to the chassis of the equipment under test, and connect the tips of the probes to the point in the circuit where measurements are to be made.

CAUTION

Never connect the ground clip of an oscilloscope probe to a circuit point other than chassis ground. The ground clip of each probe is an earth ground and will short any circuit point to which it is connected directly to earth ground. Unless that circuit is already a ground, the equipment under test can be damaged.

4. Set the vertical step attenuators of both channels as necessary to a deflection factor that allows the expected signal to be displayed without overdriving the vertical amplifier. The sensitivity/attenuation controls of both channels must be identical.
5. Set the input selector to measure ac. Operate the controls as necessary to select the **ADD** function.
6. Operate the oscilloscope controls as necessary to display a single waveform as shown in Fig. 6–14.
7. If the channel A and channel B signals are in phase, the displayed waveform is the sum of the amplitudes of the signals. Any imbalance or difference between the signals can be checked by setting the oscilloscope invert control to invert. (Typically, this inverts channel B.) The displayed waveform then becomes the difference between the two signals.
8. If the channel A and channel signals are 180° out of phase (as is the case of a push–pull amplifier shown in Fig. 6–14), it is necessary to invert one channel (channel B) by operating the oscilloscope invert control. With one channel inverted, the displayed waveform then becomes the sum of the two 180° out-of-phase signals. Any imbalance or difference between the two signals can be checked by setting the oscilloscope invert control to normal. The displayed waveform then becomes the difference between the two signals.
9. With the controls set to obtain a sum of the two signals, position the waveform as necessary to measure peak-to-peak voltage as discussed in Sec. 6–1.
10. Measure the peak-to-peak vertical deflection in centimeters. Multiply the measured distance by the vertical attenuator switch

settings. Include the attenuation factors of *both probes*. It is assumed that both probes have an identical attenuation factor.

Example: Assume a peak-to-peak vertical deflection of 6 cm (Fig. 6–14) using 10× attenuator probes and a vertical deflection factor of 20 mV/cm. Use the equation

$$\begin{matrix} \text{volts} & & \text{vertical} & & \text{calibration} & & \text{combined probe} \\ \text{peak to peak} & = & \text{deflection} & \times & \text{(mV/cm)} & \times & \text{attenuation} \\ & & \text{factor} & & \text{factor} & & \text{factors} \end{matrix}$$

Substituting the given values yields

$$\text{volts peak to peak} = 6 \times 20 \text{ mV} \times 20 = 2400 \text{ mV or } 2.4 \text{ V}$$

7

Measuring Time, Frequency, and Phase

An oscilloscope is the ideal tool for measuring time, frequency, and phase of voltages and currents. If the horizontal sweep is calibrated directly in relation to time, such as 5 ms/cm, the time duration of voltage waveforms may be measured directly on the screen without calculation. If the time duration of one complete cycle is measured, frequency can be calculated by simple division, since frequency is the reciprocal of the time duration of one cycle. If the oscilloscope is of the shop type, where the horizontal axis is not calibrated directly in relation to time, it is still possible to obtain accurate frequency and time measurements using Lissajous figures.

Measuring phase difference between two signals is equally simple with an oscilloscope. If the instrument has two vertical inputs, the signals can be displayed simultaneously and the phase measured directly on the screen. If the oscilloscope has identical vertical and horizontal amplifiers, the phase of signals can be obtained by Lissajous figures.

This chapter describes the procedures for measuring time, frequency, and phase of a-c voltages, as well as pulse and square-wave signals.

7-1. TIME DURATION MEASUREMENTS: LABORATORY OSCILLOSCOPES

NOTE

The horizontal sweep circuit of a laboratory oscilloscope is usually provided with a selector control that is direct reading in relation to time. That is, each horizontal division on the oscilloscope screen has a definite relation to time, at a given position of the horizontal sweep rate switch (such as ms/cm, μs/cm). With such oscilloscopes, the waveform can be displayed, and the time duration of the complete waveform (or any portion) can be measured directly.

1. Connect the equipment as shown in Fig. 7–1.
2. Place the oscilloscope in operation (Chapter 5).
3. Set the vertical step attenuator to a deflection factor that will allow

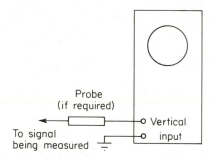

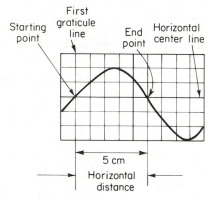

FIGURE 7-1 Measuring time duration.

the expected signal to be displayed without overdriving the vertical amplifier.

4. Connect the probe (if any) to the signal being measured.
5. Switch on the oscilloscope internal recurrent sweep. Set the horizontal sweep control to the fastest sweep rate that will display a convenient number of divisions between the time measurement points (Fig. 7-1).

NOTE

On most oscilloscopes, it is recommended that the extreme sides of the screen not be used for time-duration measurements. There may be some nonlinearity at the beginning and end of the sweep.

6. Adjust the vertical position control to move the points between which the time measurement is made to the horizontal centerline.
7. Adjust the horizontal position control to move the starting point of the time measurement area to the first graticule line.
8. Measure the horizontal distance between the time measurement points (Fig. 7-1).

NOTE

If the horizontal sweep is provided with a variable control, make certain it is off or in the calibrate position.

9. Multiply the distance measured in step 8 by the setting of the horizontal sweep control. If sweep magnification is used, divide the answer by the multiplication factor.

Example: Assume that the distance between the time measurement points is 5 cm (Fig. 7-1), the horizontal sweep control is set to 0.1 ms/cm, and there is no sweep magnification.

Use the equation

$$\text{time duration} = \frac{\text{horizontal distance (cm)} \times \text{horizontal sweep setting}}{\text{magnification}}$$

Substituting the given values yields

$$\text{time duration} = \frac{5 \times 0.1}{1} = 0.5 \text{ ms}$$

7-2. FREQUENCY MEASUREMENTS: LABORATORY OSCILLOSCOPES

NOTE

The frequency measurement of a periodically recurrent waveform is essentially the same as a time-duration measurement, except that an additional calculation must be performed. In effect, a time-duration measurement is made, then the time duration is divided into 1, or unity, since frequency of a signal is the reciprocal of one cycle.

1. Connect the equipment as shown in Fig. 7-2.
2. Place the oscilloscope in operation (Chapter 5).
3. Set the vertical step attenuator to a deflection factor that will allow the expected signal to be displayed without overdriving the vertical amplifier.
4. Connect the probe (if any) to the signal being measured.
5. Switch on the oscilloscope internal recurrent sweep. Set the horizontal sweep control to a sweep rate that will display one complete cycle of the incoming signal (Fig. 7-2).

NOTE

On most oscilloscopes, it is recommended that the extreme sides of the screen not be used for frequency measurements. There may be some nonlinearity at the beginning and end of the sweep.

6. Adjust the vertical position control so that the beginning and end points of one complete cycle are located on the horizontal centerline.

NOTE

Any two points representing one complete cycle of the waveform can be used. It is usually more convenient to measure one complete cycle at points where the waveform swings from negative to positive (or vice versa), or where the waveform starts its positive (or negative) rise. Figure 7-2 shows some typical examples of a complete cycle for various waveforms.

7. Adjust the horizontal position control to move the selected starting point of the complete cycle to the first graticule line.

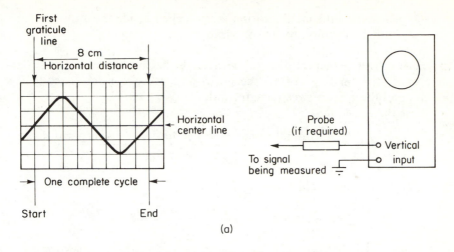

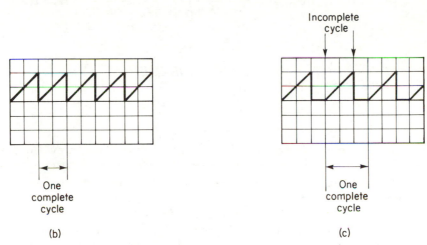

FIGURE 7-2 Measuring frequency where horizontal sweep is calibrated in relation to time.

8. Measure the horizontal distance between the beginning and end of a complete cycle.

NOTE

If the horizontal sweep is provided with a variable control, make certain it is off or in the calibrate position.

9. Multiply the distance measured in step 8 by the setting of the horizontal sweep control. If sweep magnification is used, divide the

answer by the multiplication factor. Then divide the measured time into 1, or unity, to find the frequency.

Example: Assume that the distance between the beginning and end of a complete cycle is 8 cm (Fig. 7–2a), the horizontal sweep control is set to 0.1 ms/cm, and there is no sweep magnification.
    Use the equation

$$\text{time duration} = \frac{\text{horizontal distance(cm)} \times \text{horizontal sweep setting}}{\text{magnification}}$$

Substituting the given values yields

$$\text{time duration} = \frac{8 \times 0.1}{1} = 0.8 \text{ ms}$$

$$\text{frequency} = \frac{1}{\text{time duration}} = \frac{1}{0.8} = 1250 \text{ Hz}$$

7–3.   FREQUENCY AND TIME MEASUREMENTS: SHOP OSCILLOSCOPES

NOTE

The horizontal sweep circuit of most shop oscilloscopes is provided with controls that are direct reading in relation to frequency. Usually, there are two controls: a step selector and a vernier. The sweep frequency of the horizontal trace is equal to the scale settings of the controls. Therefore, when a signal is applied to the vertical input, and the horizontal sweep controls are adjusted until one complete cycle occupies the entire length of the trace, the vertical signal is equal in frequency to the horizontal sweep control scale settings. If desired, the frequency can then be converted to time:

$$\text{time} = \frac{1}{\text{frequency}}$$

1.   Connect the equipment as shown in Fig. 7–3.
2.   Place the oscilloscope in operation (Chapter 5).
3.   Set the vertical step attenuator to a deflection factor that will allow the expected signal to be displayed without overdriving the vertical amplifier.
4.   Connect the probe (if any) to the signal being measured.

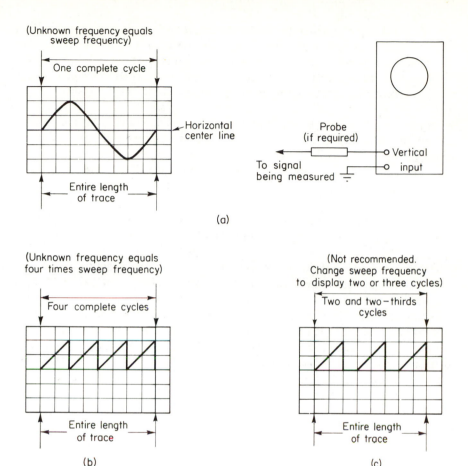

FIGURE 7-3 Measuring frequency where horizontal sweep is calibrated directly in units of frequency.

5. Switch on the oscilloscope internal recurrent sweep. Set the horizontal sweep controls (step and vernier) so that one complete cycle occupies the entire length of the trace (Fig. 7–3).

6. Read the unknown vertical signal frequency directly from the horizontal sweep frequency control settings.

Example: Assume that the step horizontal sweep control is set to the 10-kHz position, and that the vernier sweep control indicates 5 (on a total scale of 10). This indicates that the horizontal sweep frequency is 5 kHz. If one complete cycle of vertical signal occupies the entire length of the trace, the vertical signal is also at a frequency of 5 kHz.

7. If it is not practical to display only one cycle on the trace, more than one cycle can be displayed, and the resultant horizontal sweep frequency indication multiplied by the number of cycles. Two important points must be remembered: first, multiply the indicated sweep frequency by the number of cycles appearing on the trace.

Example: Assume that the step horizontal sweep control is set to the 10-kHz position, and the vernier control indicates 5 (on a total scale of 10) when three complete cycles of vertical signal occupy the entire length of the trace. This indicates that the horizontal sweep frequency is 5 kHz, and the vertical signal is at a frequency of three times that amount, or 15 kHz.

Second, it is absolutely essential that an *exact* number of cycles occupy the *entire length* of the trace.

Example: Assume that the step horizontal sweep control is set to the 10-kHz position, and the vernier control indicates 5 (on a total scale of 10) when three and one-third cycles of vertical signal occupy the entire length of the trace. This indicates that the horizontal sweep frequency is 5 kHz, and the vertical signal is at a frequency of three and one-third times that amount, or 16.5 kHz. The exact percentage of the incomplete cycle (one-third) is quite difficult to determine. It is far simpler and more accurate to increase the horizontal sweep frequency until exactly three cycles appear, or decrease the frequency until four cycles occupy the entire length of the trace.

7-4. FREQUENCY MEASUREMENTS: LISSAJOUS FIGURES

Lissajous figures or patterns can be used with almost any oscilloscope (shop or laboratory type) and will provide accurate frequency measurements. It must, however, be possible to apply an external signal to the horizontal amplifier, with the internal sweep disabled. Also, an accurately calibrated, variable-frequency signal source must be available to provide a standard frequency.

The use of Lissajous figures to measure frequency involves comparing a signal of unknown frequency (usually applied to the vertical amplifier) against a standard signal of known frequency (usually applied to the horizontal amplifier). The standard frequency is then adjusted until the pattern appears as a circle or ellipse, indicating that both signals are at the same frequency. Where it is not possible to adjust the standard signal frequency to the exact frequency of the unknown signal, the standard is adjusted to a multiple or submultiple. The pattern then appears as a number of stationary loops. The ratio of horizontal loops to vertical loops provides a measure of frequency.

1. Connect the equipment as shown in Fig. 7-4.
2. Place the oscilloscope in operation (Chapter 5).
3. Set the vertical step attenuator to a deflection factor that will allow the expected signal to be displayed without overdriving the vertical amplifier.

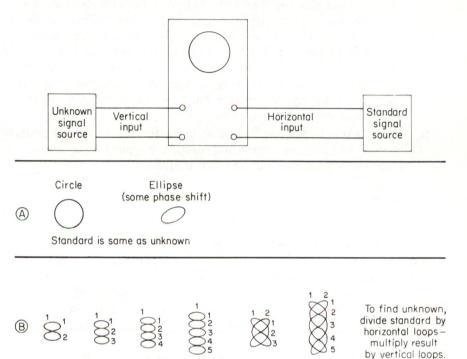

FIGURE 7-4 Measuring frequency with Lissajous patterns.

4. Switch off the oscilloscope internal recurrent sweep.

5. Set the gain controls (horizontal and vertical) to spread the pattern over as much of the screen as desired.

6. Set the position controls (horizontal and vertical) until the pattern is centered on the screen.

7. Adjust the standard signal frequency until the pattern stands still. This indicates that the standard signal is at the same frequency as the unknown frequency (if the pattern is a circle or ellipse), or that the standard signal is at a multiple or submultiple of the unknown frequency (if the pattern is composed of stationary loops). If the pattern is still moving (usually spinning), this indicates that the standard signal is not at the same frequency (or multiple) of the unknown frequency. The pattern must be stationary before the frequency can be determined.

8. Note the standard signal frequency. Using this frequency as a basis, observe the Lissajous pattern and compare it against those shown in Fig. 7–4 to determine the unknown frequency.

NOTE

If both signals are sinusoidal and are at the same frequency, the pattern will be a circle (or an ellipse, when the two signals are not exactly in phase) as shown in Fig. 7–4a.

If the standard signal (horizontal) is a multiple of the unknown signal (vertical), the pattern will show more horizontal loops than vertical loops (Fig. 7–4b). For example, if the standard signal frequency is three times that of the unknown signal frequency, there will be three horizontal loops and one vertical loop. If the standard signal frequency were 300 Hz, the unknown signal frequency would be 100 Hz. If there are two vertical loops and three horizontal loops, as another example, the unknown signal frequency would be two-thirds of the standard signal frequency.

If the standard signal (horizontal) is a submultiple of the unknown signal (vertical), the pattern will show more vertical loops than horizontal loops (Fig. 7–4c.) For example, if the standard signal frequency is one-fourth that of the unknown signal frequency, there will be four vertical loops and one horizontal loop. If the standard signal frequency were 100 Hz, the unknown signal frequency would be 400 Hz.

9. If two signals are to be matched, without regard to frequency, it is necessary to adjust only one frequency until the circle (or ellipse) pattern is obtained.

NOTE

It is recommended that the circle pattern be used for all frequency measurements whenever possible. If this is not practical, use the minimum number of loops possible. Note, too, that the use of Lissajous patterns in actual practice is quite difficult and requires considerable skill and practice to make accurate measurements.

7–5. FREQUENCY MEASUREMENTS: MODULATED RING PATTERN

In some cases, it may be difficult to use Lissajous figures, especially when there are many loops to be counted. An alternate method is use of a modulated ring. With the modulated ring method, the display appears as a wheel or gear with a number of teeth. As with Lissajous figures, the modulated ring method requires that an external signal be applied, with the internal sweep disabled, and that the external signal be an accurately calibrated, variable source to provide a standard frequency.

Using the modulated ring method involves comparing a signal of unknown frequency (usually applied to the horizontal amplifier) against the standard signal of known frequency (usually applied through a phase shift network to the vertical amplifier). The standard frequency is then adjusted until the pattern stands still and appears as a circle (or ellipse) with a number of teeth or spikes. The number of teeth indicates the frequency of the unknown signal, when multiplied by the known standard frequency.

It is essential that the unknown frequency be higher than the standard frequency to obtain a proper pattern. Also, it must be possible to increase the amplitude of the known signal above that of the unknown signal, to prevent distortion of the pattern.

1. Connect the equipment as shown in Fig. 7–5.

NOTE

The phase shift required to produce the circle (or ellipse) pattern is composed of variable resistance R_1 and fixed capacitor C_1. To obtain the correct voltage to produce a good circle, the resistance of R_1 should be equal to the reactance of C_1 at the operating frequency of the known signal source. By making R_1 variable, it is possible to match the change in C_1 reactance over a wide range of standard signal frequencies.

2. Place the oscilloscope in operation (Chapter 5).
3. Set the step attenuators to deflection factors which will allow the

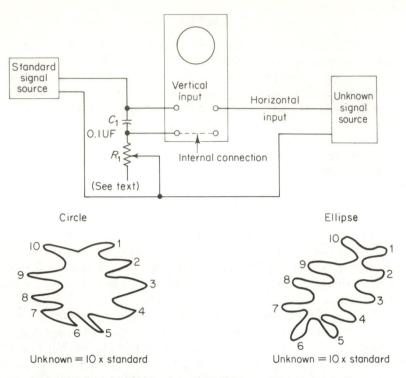

FIGURE 7-5 Measuring frequency with modulated ring pattern.

expected signals to be displayed without overdriving the amplifiers.

4. Switch off the oscilloscope internal recurrent sweep.

5. Temporarily remove the unknown signal source.

6. Set the position controls (horizontal and vertical) until the pattern is centered on the screen.

7. Adjust R_1 for a ring pattern.

8. Set the gain controls (horizontal and vertical) to spread the pattern over as much of the screen as desired.

9. Switch on the unknown signal source, and note that the ring pattern becomes modulated (teeth or spikes appear). It may be necessary to readjust the horizontal gain to produce a readable pattern.

10. Adjust the standard signal frequency until the pattern stands still. This indicates that the unknown signal is at a multiple of the known signal frequency. The pattern must be stationary before the frequency can be determined.

11. Note the standard signal frequency. Using this frequency as a

basis, observe the number of teeth appearing on the ring and determine the unknown frequency.

Example: Assume that the standard signal frequency is 1 kHz, and there are seven teeth appearing on the ring.
The unknown signal frequency is 7 kHz.

$$1 \text{ kHz} \times 7 = 7 \text{ kHz}$$

7–6.   FREQUENCY MEASUREMENTS: BROKEN RING PATTERN

The broken ring pattern is similar to the modulated ring pattern described in Sec. 7–5. Both methods are alternatives to the use of Lissajous figures for frequency measurement. With the broken ring method, the display appears as a ring broken into segments. As with the modulated ring method, the broken ring method requires that an external signal be applied, with the internal sweep disabled, and that the external signal be an accurately calibrated, variable source to provide standard frequency.

The broken ring method has an additional requirement in that the oscilloscope must be capable of Z-axis (intensity) modulation.

Use of the broken ring method involves comparing a signal of unknown frequency (applied to the Z-axis) against the standard signal of known frequency (applied to the vertical and horizontal amplifiers through a phase-shift network). The standard frequency is then adjusted until the pattern stands still and appears as a circle (or ellipse) with a number of bright traces and blanks. The bright traces or segments are made by the positive half-cycles of the unknown signal applied to the Z-axis, whereas the blanks are made by the negative half-cycles.

The number of segments (or blanks, whichever are easier to read) indicate the frequency of the unknown signal, when multiplied by the known standard frequency. It is essential that the unknown frequency be higher than the standard frequency to obtain a proper pattern.

The broken ring method is superior to the modulated ring method in that it is usually easier to distinguish blanks or traces than teeth in the ring pattern. Except for this, the accuracy of both methods is the same.

1.   Connect the equipment as shown in Fig. 7–6.

NOTE

The phase shift required to produce the circle (or ellipse) pattern is composed of variable resistance R_1 and fixed capacitor C_1. To obtain the correct voltage to produce a good circle, the resistance of

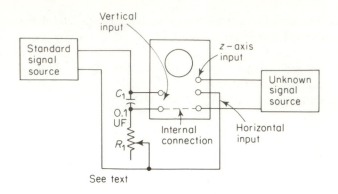

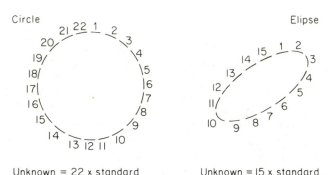

FIGURE 7-6 Measuring frequency with broken ring pattern.

R_1 should be equal to the reactance of C_1 at the operating frequency of the known signal source. By making R_1 variable, it is possible to match the change in C_1 reactance over a wide range of standard signal frequencies.

2. Place the oscilloscope in operation (Chapter 5).
3. Set the step attenuators to deflection factors that will allow the expected signals to be displayed without overdriving the amplifiers.
4. Switch off the oscilloscope internal recurrent sweep.
5. Temporarily remove the unknown signal source.
6. Set the position controls (horizontal and vertical) until the pattern is centered on the screen.
7. Adjust R_1 for a ring pattern on the screen.
8. Set the gain controls (horizontal and vertical) to spread the pattern over as much of the screen as desired.

9. Switch on the unknown signal source and note that the ring pattern is broken into segments.

10. Adjust the standard signal frequency until the pattern stands still. This indicates that the unknown signal is at a multiple of the known signal frequency. The pattern must be stationary before the frequency can be determined.

11. Note the standard signal frequency. Using this frequency as a basis, observe the number of segments (or blanks, whichever is most convenient to read) appearing on the ring and determine the unknown frequency.

Example: Assume that the standard signal frequency is 1 kHz and that there are 14 segments (or blanks) appearing on the ring.
The unknown signal frequency is 14 kHz.

$$1 \text{ kHz} \times 14 = 14 \text{ kHz}$$

7–7. FREQUENCY MEASUREMENTS: BROKEN LINE PATTERN

The broken line pattern is similar to the broken ring pattern described in Sec. 7–6. Both methods are alternatives to the use of Lissajous figures for frequency measurement. With the broken line method, the display appears as a straight horizontal line (the vertical deflection is not used) broken into segments. As with the broken ring method, the broken line method requires that an external signal be applied with the internal sweep disabled and that the external signal be an accurately calibrated, variable source to provide a standard frequency.

The broken line method requires that the oscilloscope be capable of Z-axis (intensity) modulation, but does not require a phase-shift network.

Use of the broken line method involves comparing a signal of unknown frequency (applied to the Z-axis) against the standard signal of known frequency (applied to the horizontal amplifier). The standard frequency is then adjusted until the pattern stands still and appears as a straight line with a number of bright traces and blanks. The bright traces or segments are made by the positive half-cycles of the unknown signal applied to the Z-axis, whereas the blanks are made by the negative half-cycles.

The number of segments (or blanks, whichever are easier to read) indicates the frequency of the unknown signal, when multiplied by the known standard frequency. It is essential that the unknown frequency be higher than the standard frequency to obtain a proper pattern.

The broken line method is superior to the broken ring method because it is simpler and does not require a phase-shift network, but the broken line method does not permit as high a count as the broken ring method.

1. Connect the equipment as shown in Fig. 7–7.
2. Place the oscilloscope in operation (Chapter 5).
3. Set the horizontal step attenuator to a deflection factor that will allow the expected signals to be displayed without overdriving the horizontal amplifier.
4. Switch off the oscilloscope internal recurrent sweep.
5. Temporarily remove the unknown signal source.
6. Set the position controls (horizontal and vertical) until the pattern is centered on the screen.
7. Set the horizontal gain control to spread the pattern over as much of the screen as desired.
8. Switch on the unknown signal source and note that the horizontal line is broken into segments.
9. Adjust the standard signal frequency until the pattern stands still. This indicates that the unknown signal is at a multiple of the known signal frequency. The pattern must be stationary before the frequency can be determined.
10. Note the standard signal frequency. Using this frequency as a basis, observe the number of segments (or blanks, whichever is most convenient to read) appearing on the horizontal line, and determine the unknown frequency.

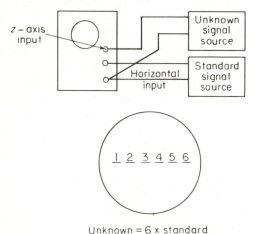

Unknown = 6 x standard

FIGURE 7-7 Measuring frequency with broken line pattern.

Example: Assume that the standard signal frequency is 1 kHz, and there are five segments (or blanks) appearing on the line.
The unknown frequency is 5 kHz.

$$1 \text{ kHz} \times 5 = 5 \text{ kHz}$$

7–8.   PHASE MEASUREMENTS: DUAL-TRACE METHOD

The dual-trace method of phase measurement provides a high degree of accuracy at all frequencies, but is especially useful at frequencies above 100 kHz where X-Y phase measurements (Sec. 7–9) may prove inaccurate owing to inherent internal phase shift.

The dual-trace method also has the advantage of measuring phase difference between signals of different amplitudes, frequency, and waveshape. The method can be applied directly to those oscilloscopes having a built-in dual-trace feature, or to a conventional single-trace oscilloscope using an electronic switch or "chopper" (Chapter 4). Either way, the procedure consists essentially of displaying both traces on the oscilloscope screen simultaneously, measuring the distance (in scale divisions) between related points on the two traces, and then converting this distance into phase.

1.   Connect the equipment as shown in Fig. 7–8.
2.   Place the oscilloscope in operation (Chapter 5).

NOTE

For the most accurate results, the cables connecting the two signals to the oscilloscope input should be of the same length and characteristics. At higher frequencies, a difference in cable length or characteristics could introduce a phase shift.

3.   Set the step attenuators to deflection factors that will allow the expected signals to be displayed without overdriving the amplifiers.
4.   Switch on the oscilloscope internal recurrent sweep.
5.   Set the position controls (horizontal and vertical) until the pattern is centered on the screen.
6.   Set the gain controls (horizontal and vertical) to spread the patterns over as much of the screen as desired.
7.   Switch on the dual-trace function of the oscilloscope, or switch on the electronic chopper.

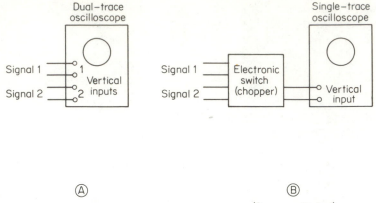

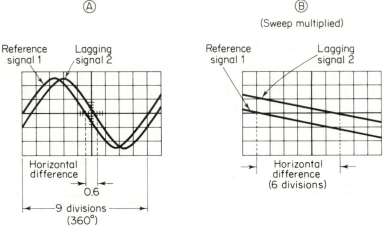

FIGURE 7-8   Measuring phase difference with dual traces.

8.   Adjust the sweep controls until one cycle of the reference signal occupies exactly nine divisions (9 cm horizontally) of the screen.

NOTE

Either of the two signals can be used as the reference signal, unless otherwise specified by requirements of the particular test. It is usually simpler if the signal of the lowest frequency is used as the reference signal.

9.   Determine the phase factor of the reference signal.

Example: If 9 cm represents one complete cycle, or 360°, then 1 cm represents 40° [360° ÷ 9 divisions (cm) = 40°/cm].

10. Measure the horizontal distance between corresponding points on the waveform. Multiply the measured distance (in centimeters) by $40°$ (phase factor) to obtain the exact amount of phase difference.

Example: Assume a horizontal difference of 0.6 cm with a phase factor of $40°$ as shown in Fig. 7–8a.
Use the equation

$$\text{phase difference} = \begin{array}{c} \text{horizontal} \\ \text{difference} \\ \text{(cm)} \end{array} \times \begin{array}{c} \text{phase} \\ \text{factor} \end{array}$$

Substituting the given values yields

$$\text{phase difference} = 0.6 \times 40°$$

The phase difference would be $24°$.

11. If the oscilloscope is provided with a sweep magnification control where the sweep rate is increased by a fixed amount ($5\times$, $10\times$, etc.), and only a portion of one cycle can be displayed, more accurate phase measurements can be made. In this case, the phase factor is determined as described in step 9. Then the approximate phase difference is determined as described in step 10. Without changing any other controls, the sweep rate is increased (by the sweep magnification control or the sweep rate control) and a new horizontal distance measurement is made, as shown in Fig. 7–8b.

Example: If the sweep rate were increased 10 times (with the magnifier or sweep rate control), the adjusted phase factor would be $40° \div 10 = 4°/\text{cm}$. Figure 7–8b shows the same signal as used in Fig. 7–8a, but the sweep rate set to $\times 10$. With a horizontal difference of 6 cm, the phase difference would be

$$\text{phase difference} = \begin{array}{c} \text{horizontal} \\ \text{difference} \\ \text{(cm)} \end{array} \times \begin{array}{c} \text{adjusted} \\ \text{phase} \\ \text{factor} \end{array}$$

Substituting the given values yields

$$\text{phase difference} = 6 \times 4°$$

The phase difference would be $24°$.

7-9. PHASE MEASUREMENTS:
X-Y METHOD

The *X-Y* phase measurement method can be used to measure the phase difference between two sine-wave signals of the *same frequency*. This method provides a method of measurement for signal frequencies up to about 100 kHz more precise than the dual-trace method discussed in Sec. 7-8. Above this frequency, however, the inherent phase difference between the horizontal and vertical systems makes accurate phase measurement difficult. Therefore, the *X-Y* method should be limited to phase measurement of lower-frequency signals and to signals of the same frequency.

In the *X-Y* method, one of the sine-wave signals provides horizontal deflection (*X*), and the other provides the vertical deflection (*Y*). The phase angle between the two signals can be determined from the resulting Lissajous pattern.

1. Connect the equipment as shown in Fig. 7-9a.

NOTE

Figure 7-9a shows the test connection necessary to determine the inherent phase shift (if any) between the horizontal and vertical

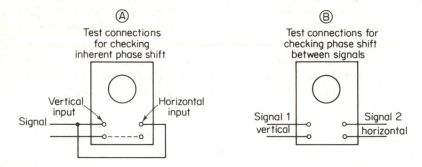

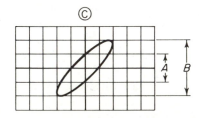

FIGURE 7-9 Measuring phase difference with *X-Y* method.

deflection systems of the oscilloscope. Even the most expensive laboratory oscilloscopes with identical horizontal and vertical amplifiers will have some inherent phase shift, particularly at the higher frequencies. Therefore, all oscilloscopes should be checked and the inherent phase shift recorded before any phase measurements are made. Inherent phase shift also should be checked periodically. If there is excessive phase shift (in relation to the anticipated phase shift of signals to be measured), the oscilloscope should not be used. A possible exception exists when the signals to be measured are of sufficient amplitude to be applied directly to the oscilloscope deflection plates, and thus bypass the horizontal and vertical amplifiers.

2. Place the oscilloscope in operation (Chapter 5).
3. Set the step attenuators to deflection factors that will allow the expected signals to be displayed without overdriving the amplifiers.
4. Switch off the oscilloscope internal recurrent sweep.
5. Set the gain controls (horizontal and vertical) to spread the pattern over as much of the screen as desired.
6. Set the position controls (horizontal and vertical) until the pattern is centered on the screen. Center the display in relation to the vertical graticule line. Measure distance *A* and *B*, as shown in Fig. 7–9c. Distance *A* is the vertical measurement between the two points where the trace crosses the vertical centerline. Distance *B* is the maximum vertical height of the display.
7. Divide *A* by *B* to obtain the sine of the phase angle between the two signals. The angle can then be obtained from Table 7–1. The resultant angle is the inherent phase shift.

NOTE

If the display appears as a diagonal straight line, the two amplifiers are either in phase (tilted upper right to lower left) or 180° out of phase (tilted upper left to lower right). If the display is a circle, the signals are 90° out of phase. Figure 7–10 shows the Lissajous displays produced between 0° and 360°. Notice that above 180° phase shift, the resultant display will be the same as at some lower frequency. Therefore, it may be difficult to tell whether the signal is leading or lagging. One way to determine correct phase (leading or lagging) is to introduce a small, known phase shift to one of the inputs. The proper angle may then be determined by noting the direction in which the pattern changes.

8. Once the inherent phase shift has been determined, connect the

TABLE 7-1. Table of Sines

Sine	Angle	Sine	Angle
0.0000	0	0.7193	46
0.0175	1	0.7314	47
0.0349	2	0.7431	48
0.0523	3	0.7547	49
0.0689	4	0.7660	50
0.0872	5	0.7771	51
0.1045	6	0.7880	52
0.1219	7	0.7986	53
0.1392	8	0.8090	54
0.1564	9	0.8192	55
0.1736	10	0.8290	56
0.1908	11	0.8387	57
0.2079	12	0.8480	58
0.2250	13	0.8572	59
0.2419	14	0.8660	60
0.2588	15	0.8746	61
0.2756	16	0.8829	62
0.2924	17	0.8910	63
0.3090	18	0.8988	64
0.3256	19	0.9063	65
0.3420	20	0.9135	66
0.3584	21	0.9205	67
0.3746	22	0.9272	68
0.3907	23	0.9336	69
0.4067	24	0.9397	70
0.4226	25	0.9455	71
0.4384	26	0.9511	72
0.4540	27	0.9563	73
0.4695	28	0.9613	74
0.4848	29	0.9659	75
0.5000	30	0.9703	76
0.5150	31	0.9744	77
0.5299	32	0.9781	78
0.5446	33	0.9816	79
0.5592	34	0.9848	80
0.5736	35	0.9877	81
0.5878	36	0.9903	82
0.6018	37	0.9925	83
0.6157	38	0.9945	84
0.6293	39	0.9962	85
0.6428	40	0.9976	86
0.6561	41	0.9986	87
0.6691	42	0.9994	88
0.6820	43	0.9998	89
0.6947	44	1.0000	90
0.7071	45		

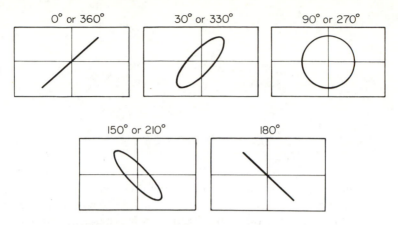

FIGURE 7-10 Phase of typical *X-Y* displays.

equipment as shown in Fig. 7–9b. Repeat steps 3 through 7 to find the phase angle between the two signals.

9. Subtract the inherent phase difference from the phase angle to determine the true phase difference.

Example: Assume an inherent phase difference of 2°, with a display as shown in Fig. 7–9c, where *A* is 2 cm and *B* is 4 cm.

Use the equation

$$\text{sine of phase angle} = \frac{A}{B}$$

Substituting the given values yields

$$\text{sine of phase angle} = \frac{2}{4} = 0.5$$

From Table 7–1,

$$\text{sine of phase angle} = 30°$$

To adjust for the phase difference between *X* and *Y* amplifiers, subtract the inherent phase factor

$$\begin{matrix} \text{actual} & & \text{inherent} \\ \text{phase} & = \text{sine of phase angle} - & \text{phase} \\ \text{factor} & & \text{difference} \end{matrix}$$

Substituting the given values yields

$$\begin{aligned}\text{actual} \\ \text{phase} &= 30° - 2° = 28° \\ \text{factor}\end{aligned}$$

7-10. PHASE MEASUREMENTS BETWEEN VOLTAGE AND CURRENT

It is sometimes necessary to measure the phase difference between a voltage and current applied across the same load. This can be accomplished by passing a portion of the current through a fixed resistor, thus converting the current to a voltage. The phase of the resultant voltage is then compared to the load voltage phase. A resistor, capable of the necessary wattage dissipation, is the only other component required for the procedure. Either the X-Y method or the dual-trace method can be used for the actual phase comparison.

Figure 7–11 shows the test connections required for converting the current into a voltage and applying both voltages to the oscilloscope.

In Fig. 7–11a, the signal voltage E_1 is applied across the load and test resistor R_1. Voltage E_1 is also applied to one of the vertical inputs. The current-developed voltage E_2 appears across R_1, and is applied to the other vertical input.

In Fig. 7–11b, the signal voltage E_1 is applied across the load and test resistor R_1. Voltage E_1 is also applied to the electronic switch (chopper). The current-developed voltage E_2 appears across R_1 and is applied to the other electronic switch input.

In Fig. 7–11c, the signal voltage E_1 is applied across the load and test resistor R_1. Voltage E_1 is also applied to the vertical input. The current-developed voltage E_2 appears across R_1 and is applied to the horizontal input.

Once the test connections have been made, the phase difference between voltage and current can be determined by the procedures of Sec. 7–8 (dual trace), or Sec. 7–9 (X-Y method), whichever applies.

NOTE

The actual resistance value of R_1 is not critical. It should be low in comparison to the resistance value of the load. Usually, 1 to 10 Ω is adequate to develop sufficient voltage for measurement. The wattage of the resistor R_1 must be at least double the square of the maximum current (in amperes). For example, if the maximum anticipated current is 10 A, the minimum wattage of the resistor should be $10^2 \times 2 = 200$ W.

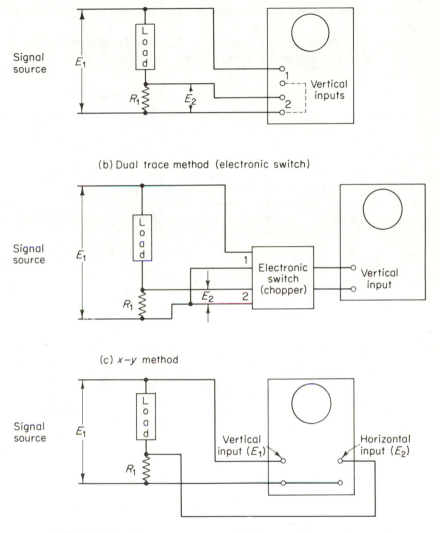

(a) Dual trace method (dual trace oscilloscope)

Signal source

E_1

Load

R_1

E_2

1

2

Vertical inputs

(b) Dual trace method (electronic switch)

Signal source

E_1

Load

R_1

E_2

1

2

Electronic switch (chopper)

Vertical input

(c) x–y method

Signal source

E_1

Load

R_1

Vertical input (E_1)

Horizontal input (E_2)

FIGURE 7-11 Phase measurement between voltage and current.

7-11. PHASE MEASUREMENTS BETWEEN TWO CURRENTS

It is sometimes necessary to measure the phase difference between two currents. This can be accomplished by passing the currents through fixed resistors, thus converting the currents into voltages. The phase of the resultant voltages is then compared. Two resistors, capable of the necessary wattage dissipation, are the only other components required for the procedure. Either the *X-Y* or the dual-trace method can be used for the actual phase comparison.

Figure 7–12 shows the test connections required for converting the currents into voltages and applying both voltages to the oscilloscope.

(a) Dual trace method (dual trace oscilloscope)

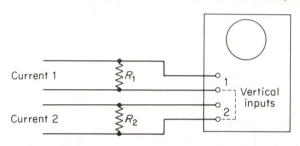

(b) Dual trace method (electronic switch)

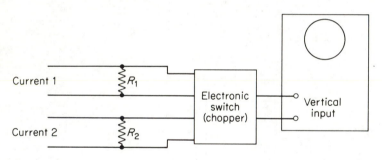

(c) *x–y* method

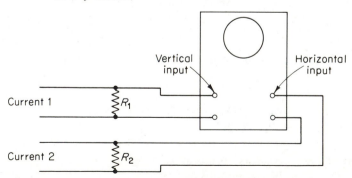

FIGURE 7-12   Phase measurements between currents.

In Fig. 7–12a, the currents are applied across corresponding resistors R_1 and R_2, with the resulting voltages applied to the two corresponding vertical inputs.

In Fig. 7–12b, the currents are applied across corresponding resistors R_1 and R_2, with the resulting voltages applied to the two corresponding electronic switch inputs.

In Fig. 7–12c, the currents are applied across corresponding resistors R_1 and R_2, with the resulting voltages applied to the vertical and horizontal inputs.

Once the test connections have been made, the phase difference between the two currents can be determined by the procedures of Sec. 7–8 (dual trace) or Sec. 7–9 (*X-Y* method), whichever applies.

NOTE

The actual resistance value of R_1 is not critical. It should, however, be low in comparison to the load. Usually, 1 to 10 Ω is adequate to develop sufficient voltage for measurement. The wattage of each resistor must be at least double the square of the maximum current (in amperes). For example, if the maximum anticipated current is 10 A, the minimum wattage of the resistor should be $10^2 \times 2 = 200$ W.

8

Using Oscilloscopes with Sweep Generators

The prime function of a sweep generator is the sweep frequency alignment of TV and FM receivers. In this application, sweep generators are used with oscilloscopes to display the bandpass characteristics of the receiver under test. The sweep generator/oscilloscope combination can also be used effectively to check the operation of filters, to check impedance of such items as transmission lines, antennas, and tuning stubs, and to determine impedance match.

A sweep generator is an FM generator. When a sweep generator is set to a given frequency, this is a center frequency. In essence, a sweep generator is a frequency-modulated radio-frequency generator. The usual frequency modulation rate is 60 Hz for most TV and FM sweep generators. Other sweep rates could be used, but since power lines usually have a 60-Hz frequency, this frequency is both convenient and economical for the sweep rate.

Some sweep generators incorporate a marker generator. Marker signals are necessary to pinpoint frequencies when making sweep frequency alignments and tests. Although sweep generators are accurate in both center frequency and sweep width, it is almost impossible to pick out a particular frequency along the spectrum of frequencies being swept. Therefore, fixed-frequency "marker" signals are injected into the circuit together with the sweep frequency generator output. On sweep generators without a built-in marker generator, markers can be added by means of an absorption-type marker adder. These marker adders can be built in, or they are available as accessories. Where a marker adder would provide too limited a number of fixed frequency points, a marker generator can be used in conjunction with a sweep generator and an oscilloscope. Basically, a marker generator is an RF signal generator which has highly accurate dial markings, and which can be calibrated precisely against internal or external signals. The sweep generator is tuned to sweep the band of frequencies passed by the wideband circuits (tuner, IF, video, filter, etc.), and a trace representing the response characteristics of the circuits is displayed on the oscilloscope. The marker generator is used to provide calibrated markers along the response curve. When the marker signal from the marker generator is coupled into the test circuit, a vertical "pip," or marker, appears on the curve. When the marker generator is tuned to a frequency within the passband accepted by the equipment under test, the marker indicates the position of that frequency on the sweep trace.

Another feature found on some sweep generators is a blanking circuit. When the sweep generator output is swept across its spectrum, the frequencies go from low to high, then return from high to low. With the blanking circuit actuated, the return or retrace is blanked off. This makes it possible to view a zero-reference line on the oscilloscope during the retrace period.

8–1. BASIC SWEEP GENERATOR/OSCILLOSCOPE TEST PROCEDURE

The following steps describe the *basic* procedure for using a sweep generator with an oscilloscope. Later sections in this chapter describe procedures for using the sweep generator/oscilloscope combination to test specific equipment or circuits.

1. Connect the equipment as shown in Fig. 8–1.
2. Place the oscilloscope in operation (Chapter 5).
3. Place the sweep generator in operation as described in its instruction manual.
4. Switch off the oscilloscope internal recurrent sweep.
5. Set the oscilloscope sweep selector and sync selector to external. Under these conditions, the oscilloscope horizontal sweep should be obtained from the generator sweep output, and the length of the horizontal sweep should represent total sweep spectrum. For example, if the sweep is from 10 to 20 kHz, the left-hand end of the horizontal trace represents 10 kHz and the right-hand end represents 20 kHz. Any point along the horizontal trace will represent a

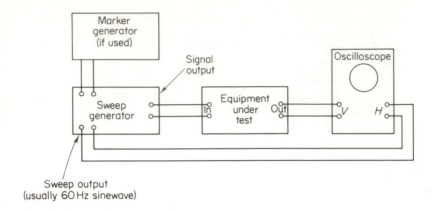

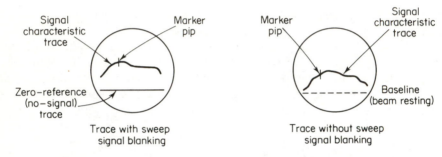

FIGURE 8-1 Basic sweep generator/oscilloscope test connections.

corresponding frequency. For example, the midpoint on the trace would represent 15 kHz. If a rough approximation of frequency is desired, the horizontal gain control can be adjusted until the trace occupies an exact number of scale divisions, such as 10 cm for the 10-20-kHz sweep signal. Each centimeter division would then represent 1 kHz.

6. If a more accurate frequency measurement is desired, the marker generator must be used. The marker generator output frequency is adjusted until the marker pip is aligned at the desired point on the trace. The frequency is then read from the marker generator frequency dial.

7. The response curve (trace) depends upon the device under test. If the device has a passband (as do most receiver circuits) and the sweep generator is set so that its sweep is wider than the passband, the trace will start low at the left, rise toward the middle, and then drop off at the right, as shown in Fig. 8–1. The sweep generator/oscilloscope method will tell at a glance the overall passband characteristics of the device (sharp response, flat response, irregular

response at certain frequencies, etc.). The exact frequency limits of the passband can be measured with the marker generator pip.

8. Switch the sweep generator blanking control on or off as desired. Some sweep generators do not have a blanking function. With the blanking function in effect, there will be a zero-reference line on the trace. With the blanking function off, the horizontal baseline will not appear. The sweep generator blanking function is not to be confused with the oscilloscope blanking (which is bypassed when the sweep signal is applied directly to the horizontal amplifier).

8–2. CHECKING SWEEP GENERATOR OUTPUT UNIFORMITY

Many sweep generators do not have a uniform output. That is, the output voltage is not constant over the swept band. In some tests, this can lead to false conclusions. In other tests, it is only necessary to know the amount of nonuniformity, and then make allowances. For example, a sweep generator can be checked for flatness before connection to a circuit, and any variation in output noted. If the output remains the same after it is connected to the circuit, even though it may have variations, the circuit under test is not at fault. Sweep generator output can be checked as follows:

1. Connect the equipment as shown in Fig. 8–2.

NOTE

The RF or demodulator probe shown in Fig. 8–2 can be omitted if the sweep generator output signal frequency is within the passband of the oscilloscope vertical amplifier. This is not usually the case with shop-type oscilloscopes, so some form of demodulator is necessary. If the oscilloscope is not equipped with a demodulator or RF probe, demodulator networks can be fabricated as shown in Fig. 8–3. Figure 8–3a is for a single-ended sweep generator output; Fig. 8–3b is for a double-ended or balanced output. The balanced output is often found on sweep generators designed specifically for use with TV and FM receivers.

2. Place the oscilloscope in operation (Chapter 5).
3. Place the sweep generator in operation as described in its instruction manual. Set sweep width to maximum.
4. Switch off the oscilloscope internal recurrent sweep.
5. Set the oscilloscope sweep selector and sync selector to external so

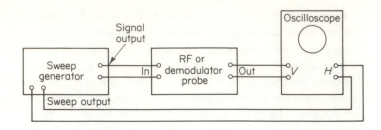

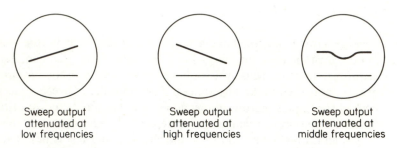

FIGURE 8-2 Checking sweep generator output uniformity.

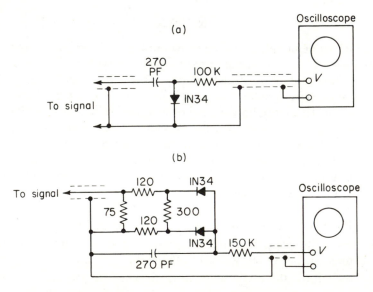

FIGURE 8-3 RF demodulator probe circuits.

that the horizontal sweep is obtained from the generator sweep output.

6. Switch the sweep generator blanking control on or off as desired. With the blanking function not in effect, only one trace will appear. This should be deflected vertically from the normal trace resting position. With the blanking function on, there will be two traces. The upper trace is the generator output characteristic; the lower trace is the no-signal trace.

7. Check the sweep generator output for flatness. If the right-hand side of the trace drops or slopes, this indicates that the sweep output is reduced at the high-frequency end of the sweep. A slope to the left indicates a reduced output at the low-frequency end of the sweep. If the trace drops off suddenly, or dips in the middle, this indicates an uneven output. A perfectly flat trace or (more realistically) a trace that has only a slight curvature at the ends indicates an even output across the entire swept band.

8. Leave the sweep width at maximum, but adjust the center frequency of the sweep generator over its entire range. Check that the output is flat, or at least that any variations are consistent, across the range of the sweep generator.

8-3. ALTERNATE SWEEP GENERATOR OPERATION

The procedures described in this chapter can be performed with the oscilloscope sweep selector set to internal, and the sync selector set to line. Two conditions must be met. First, the sweep generator must be swept at the line frequency. Second, the oscilloscope, or sweep generator, must have a phasing control so that the two sweeps can be synchronized. The phasing problem is discussed in Sec. 2-3. This alternate method is used where the sweep generator does not have a sweep output separate from the signal output, or when it is not desired to use the sweep output. Blanking of the trace (if any blanking is used) is controlled by the oscilloscope circuits.

8-4. MEASURING TRANSMISSION-LINE IMPEDANCE

The sweep generator/oscilloscope combination can be used to measure the impedance of transmission lines, TV antenna lead-in wire, or cables. This method is particularly effective since it provides an impedance check over a broad frequency range.

1. Connect the equipment as shown in Fig. 8–4.

2. Place the oscilloscope in operation (Chapter 5). Switch off internal recurrent sweep. Set the oscilloscope sweep selector and sync selector to external.

3. Place the sweep generator in operation as described in its instruction manual. Switch the sweep generator blanking control on or off as desired. Tune the sweep generator to the normal operating frequency with which the transmission line is to be used. Adjust the sweep width to cover the complete range of frequencies.

4. Adjust the variable resistance to the supposed impedance value of the transmission line.

5. The oscilloscope pattern should show a flat trace. If not, adjust the variable resistance until the trace is flat.

6. Disconnect the variable resistance (without disturbing its setting) and measure the resistance with an ohmmeter. This value is equal to the transmission line characteristic impedance.

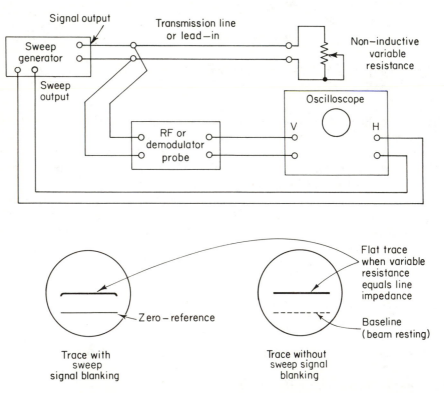

FIGURE 8-4 Measuring transmission-line impedance.

7. If the trace cannot be made flat, temporarily disconnect the transmission line but leave the sweep generator output connected to the oscilloscope. Check the trace pattern. If it is still not flat, the sweep generator output is not uniform (refer to Sec. 8–2).

8–5. MEASURING IMPEDANCE MATCH

The sweep generator/oscilloscope combination can be used to measure impedance match between a transmission line and antenna or any other terminating device. This method provides a check of impedance match over a broad frequency range.

1. Connect the equipment as shown in Fig. 8–5.

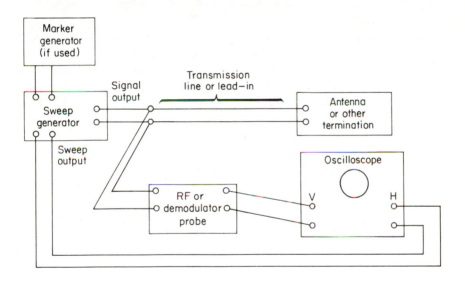

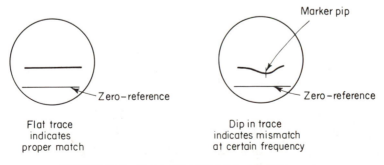

FIGURE 8-5 Measuring impedance match.

2. Place the oscilloscope in operation (Chapter 5). Switch off the internal recurrent sweep. Set the oscilloscope sweep selector and sync selector to external.

3. Place the sweep generator in operation as described in its instruction manual. Switch the sweep generator blanking control on or off as desired. Tune the sweep generator to the normal operating frequency with which the transmission line and antenna (or other terminating device) are to be used. Adjust the sweep width to cover the complete range of frequencies.

4. If the transmission line and antenna are properly matched, the oscilloscope will show a flat trace. If the oscilloscope trace is not flat, and it is desired to determine the frequency at which the mismatch occurs, the marker generator can be adjusted until the marker pip is aligned at the desired point on the trace. The mismatch frequency, or band of frequencies, can be read from the marker generator frequency dial.

5. If there is any doubt that the variation in trace flatness is caused by variation in sweep generator output, temporarily disconnect the transmission line but leave the sweep generator output connected to the oscilloscope. If the variation is removed, or reduced drastically when the transmission line and antenna (or other termination) are disconnected, this indicates a mismatch.

8-6. CHECKING RESONANT FREQUENCY OF ANTENNA TUNING DEVICES

Often, antennas used with transmission lines are tuned by an external device. An example of this is TV antennas which are often tuned to cover a specific frequency by means of a stub. The sweep generator/oscilloscope combination can be used to measure the resonant frequency of the tuning device.

NOTE

This same basic procedure can be used to tune (or to check the resonant frequency) of any resonant circuit. It is essential, however, that the sweep generator output be at the fundamental frequency of the resonant circuit for accurate results.

1. Connect the equipment as shown in Fig. 8-6.

2. Place the oscilloscope in operation (Chapter 5). Switch off the internal recurrent sweep. Set the oscilloscope sweep selector and sync selector to external.

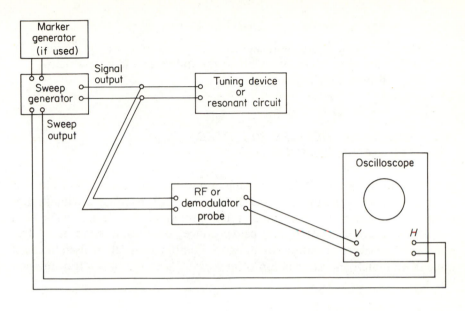

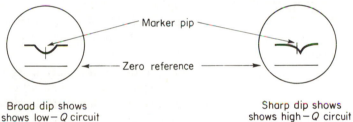

Marker pip

Zero reference

Broad dip shows
shows low − Q circuit

Sharp dip shows
shows high − Q circuit

FIGURE 8-6 Checking resonant frequency of antenna tuning devices.

3. Place the sweep generator in operation as described in its instruction manual. Switch the sweep generator blanking control on or off as desired. Tune the sweep generator to the frequency at which the tuning device is supposed to be resonant. Adjust the sweep width to cover a wide range of frequencies, but not so wide that the nearest harmonic is covered. If harmonics are covered, their indications on the trace may prove confusing.

4. Note the point on the trace at which the dip occurs. The dip indicates the frequency of the resonant circuit or tuning device.

5. The procedure can be reversed to adjust the tuning device to a given frequency, if desired.

6. For greatest accuracy, adjust the marker generator until the marker pip is aligned at the center of the trace dip. The exact resonant frequency can then be read from the marker generator frequency dial.

NOTE

The dip characteristics indicate the sharpness or Q of the tuning device or resonant circuit. A broad dip indicates a low Q; a sharp dip indicates a high Q.

8-7. CHECKING DEVICES INSERTED IN TRANSMISSION LINES

When any device is inserted in a transmission line, there occurs a possibility of mismatch between the line and the device. This will produce attenuation at certain frequencies. Even if there is no mismatch, the device (such as a coupler, splitter, pick-off probe) can produce some attenuation at all frequencies. The sweep generator/oscilloscope combination can be used to check impedance match and attenuation of any device inserted into a transmission line.

1. Connect the equipment as shown in Fig. 8-7.
2. Place the oscilloscope in operation (Chapter 5). Switch off the internal recurrent sweep. Set the oscilloscope sweep selector and sync selector to external.
3. Place the sweep generator in operation as described in its instruction manual. Switch the sweep generator blanking control on or off as desired. Tune the sweep generator to the normal operating frequency with which the transmission line and inserted device are to be used. Adjust the sweep width to cover the complete range of frequencies.
4. With the probe connected at the sweep generator output, the oscilloscope will show a flat trace if the transmission line and inserted device are properly matched. If the trace is not flat, and it is desired to determine the frequency at which the mismatch occurs, the marker generator can be adjusted until the marker pip is aligned at the desired point on the trace. The mismatch frequency, or band of frequencies, can be read from the marker generator frequency dial.
5. If there is any doubt that the variation in trace flatness is caused by variation in sweep generator output, temporarily disconnect the transmission line, but leave the oscilloscope probe connected to the sweep generator output. If the variation is removed, or drastically reduced when the transmission line and inserted device are disconnected, this indicates a mismatch.
6. Once the match or mismatch has been established, connect the probe to the inserted device input at point A (Fig. 8-7). This will

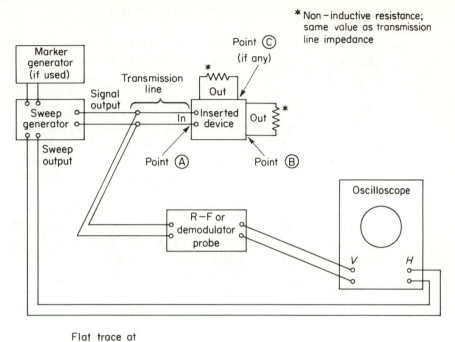

FIGURE 8-7 Checking devices inserted in transmission lines.

establish a vertical reference deflection. Then connect the probe to point *B*, and to point *C* (if any) in turn.

7. The oscilloscope vertical deflection should be lower at points *B* and *C* than at point *A* because almost any device inserted into a transmission line will show some loss or attenuation. Unless there is some special design characteristic, the output at points *B* and *C* should be the same. The voltage difference between the signal at points *A* and *B* (or *A* and *C*) can be measured directly on the oscilloscope (assuming that the vertical system is voltage calibrated) and converted to a ratio, decibel value, or whatever is desired.

8-8.   MEASURING INPUT AND OUTPUT IMPEDANCES

The input and output impedances of RF (and audio) components can be checked quickly and over a broad range of frequencies with the sweep generator/oscilloscope combination.

1.   Connect the equipment as shown in Fig. 8–8.

2.   Place the oscilloscope in operation (Chapter 5). Switch off the internal recurrent sweep. Set the oscilloscope sweep selector and sync selector to external.

3.   Place the sweep generator in operation as described in its instruction manual. Switch the sweep generator blanking control on or off as desired. Tune the sweep generator to the normal operating frequency with which the component is to be used. Adjust the sweep width to cover the complete range of frequencies.

4.   Disconnect resistors R_1 and R_2 from the circuit.

5.   Adjust the sweep generator output level so that the trace is at a convenient vertical scale marking. Note this vertical deflection scale marking.

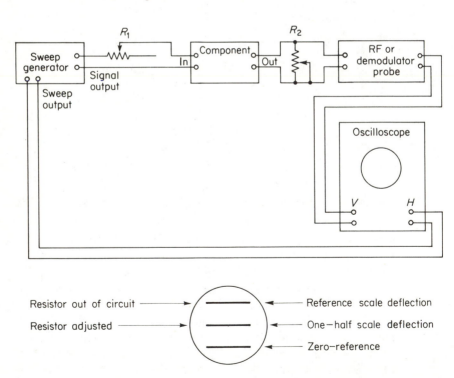

FIGURE 8-8   Measuring input and output impedances.

6. Connect resistor R_1 back into the circuit. Vary the resistance of R_1 until the voltage indicated on the oscilloscope is one-half the original value.

7. Disconnect resistor R_1 from the circuit, and measure its d-c resistance. This resistance is equivalent to the input impedance of the component.

8. With both R_1 and R_2 out of the circuit, again adjust the sweep generator output level so that the trace is at a convenient vertical scale indication.

9. Connect resistor R_2 back into the circuit. Vary the resistance of R_2 until the voltage indication on the oscilloscope is one-half the original value.

10. Disconnect resistor R_2 from the circuit and measure its d-c resistance. This resistance is equivalent to the output impedance of the component.

NOTE

This method is accurate if the impedance to be measured is *resistive,* but only approximate if the impedance is reactive.

8–9. MEASURING INPUT IMPEDANCE

Many RF components have an input circuit, but no measurable output. An antenna is a good example of such a component. The input impedance of these components can be measured quickly and over a broad range of frequencies with the sweep generator/oscilloscope combination.

1. Connect the equipment as shown in Fig. 8–9.

2. Place the oscilloscope in operation (Chapter 5). Switch off the internal recurrent sweep. Set the oscilloscope sweep selector and sync selector to external.

3. Place the sweep generator in operation as described in its instruction manual. Switch the sweep generator blanking control on or off as desired. Tune the sweep generator to the normal operating frequency with which the component is to be used. Adjust the sweep width to cover the complete range of frequencies.

4. Set switch S_1 to the A position.

5. Adjust the sweep generator output level so that the trace is set at a convenient vertical scale marking. Note this vertical deflection scale marking.

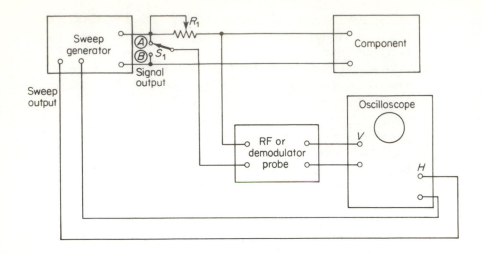

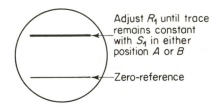

Adjust R_1 until trace remains constant with S_1 in either position A or B

Zero-reference

FIGURE 8-9 Measuring input impedance.

6. Move the test switch S_1 from position A to position B.

7. Adjust resistor R_1 until the voltage indicated on the oscilloscope is the same in both positions of switch S_1.

8. Disconnect resistor R_1 from the circuit and measure its d-c resistance. This resistance is equivalent to the input impedance of the component.

NOTE

This method is accurate if the impedance to be measured is *resistive*, but only approximate if the impedance is reactive.

8-10. CHECKING AUDIO FILTERS

The response characteristics of audio filters can be checked using the sweep generator/oscilloscope combination, provided that the sweep generator is capable of sweeping over the audio range (from approximately 20 Hz to 20 kHz). If maximum accuracy is desired, an audio oscillator must also be used as a marker generator.

1.   Connect the equipment as shown in Fig. 8–10.

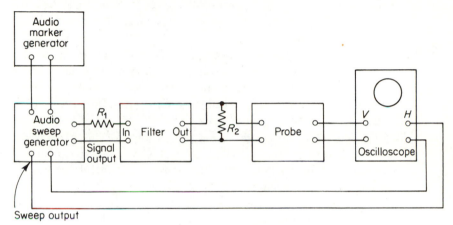

R_1 = filter input impedance
R_2 = filter output impedance

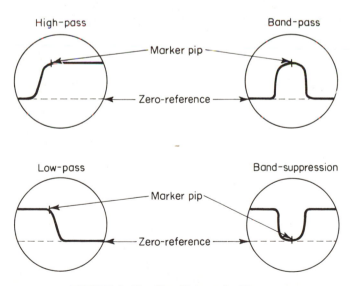

FIGURE 8–10 Checking audio filters.

NOTE

Resistors R_1 and R_2 are included since many test specifications for filters require that the input and output be terminated in their respective impedances. R_1 and R_2 may be omitted unless required by specification.

2. Place the oscilloscope in operation (Chapter 5). Switch off the internal recurrent sweep. Set the oscilloscope sweep selector and sync selector to external.

3. Place the sweep generator in operation as described in its instruction manual. Switch the sweep generator blanking control on or off as desired. Adjust the sweep generator to cover the complete audio range, or that portion of the range which would affect filter operation.

4. Check the filter response curve appearing on the oscilloscope against those of Fig. 8–10, or against the filter specifications. Typical high-pass, low-pass, band-pass, and band-suppression response curves are shown in Fig. 8–10.

5. If it is desired to determine the exact frequencies at which filter response occurs, the audio marker generator can be adjusted until the marker pip is aligned at the point of interest. The frequency, or band of frequencies, can be read from the audio marker frequency dial.

6. The amplitude of any point on the response curve can be measured directly on the oscilloscope (assuming that the vertical system is voltage-calibrated).

8-11. SWEEP FREQUENCY ALIGNMENT OF RECEIVERS

As stated in Sec. 8–1, the sweep generator/oscilloscope combination can be used most effectively in sweep alignment of receivers. Procedures for sweep generator/oscilloscope alignment of AM and FM communications receivers are given in Chapter 11. Chapter 13 gives procedures for sweep alignment of TV receivers.

9

Checking Individual Components

An oscilloscope is particularly useful in checking those components where response curves, transient characteristics, phase relationships, and time are of special importance. Of course, any component that can be checked by a voltmeter can also be checked with an oscilloscope, since the oscilloscope can function as an a-c or d-c voltmeter. If a component requires only a voltage check (or resistance check) it is more economical and practical to use a simple multitester.

This chapter describes the procedures for checking those components where an oscilloscope can provide a better or faster method. In most cases, similar tests can be made without an oscilloscope, but that involves making a series of many tests, and then plotting the data as a curve or graph. The oscilloscope method presents the information instantly.

9-1. TESTING TRANSISTORS FOR COLLECTOR CURRENT VERSUS EMITTER CURRENT USING A CONVENTIONAL OSCILLOSCOPE

A d-c oscilloscope can be used to display and measure the collector-current versus the emitter-current characteristics of transistors. Both the vertical and horizontal channels must be voltage-calibrated. The procedures for voltage calibration of the vertical channel are given in Chapter 6. The same procedures can be applied to the horizontal channel, so that a volts/division calibration is obtained instead of the usual time/division. The horizontal zero reference point should, however, be at the left (or right) of the screen rather than in the center. Also, the horizontal and vertical channels must be identical, or nearly identical, in order to eliminate any phase difference. Section 7-9 describes the procedures for determining the phase difference between the horizontal and vertical channels.

As shown in Fig. 9-1, the transistor is tested by applying a controlled d-c voltage to the collector. The collector voltage is developed by rectifying the transformer T_1 secondary voltage with diode CR_1. Collector voltage can be adjusted to any desired value by the variac. When collector current flows on positive half-cycles, the current flows through resistor R_1. The voltage drop across R_1 is applied to the vertical channel and causes the spot to move up and down. Therefore, vertical deflection is proportional to current. The vertical scale divisions can be converted directly to current (in amperes) when R_1 is made 1 Ω. With R_1 at a value of 10 Ω, the indicated voltage value must be divided by 10 to obtain current. For example, a 3-V vertical deflection indicates a 0.3-A current.

The same voltage applied to the transistor collector is applied to the horizontal channel (which has been voltage-calibrated), and causes the spot to move from left to right (for the NPN transistor shown). Therefore, horizontal deflection is proportional to voltage. The combination of the horizontal (voltage) deflection and vertical (current) deflection causes the spot to trace out the collector-current versus collector-voltage characteristics of the transistor.

Usually, the change in collector current for a given change in emitter current is the desired characteristic for most transistors. This can be displayed by setting the emitter current to a given value and measuring the collector-current curve, with a given collector voltage. Then the emitter current is changed to another value, and the new collector current is displayed, without changing the collector voltage. Collector voltage is set by the variac. Emitter current is set by R_2 and measured by M_1.

The test connection diagram of Fig. 9-1 is for an NPN transistor connected in a common emitter circuit. If a PNP transistor is to be tested, the polarity of rectifier CR_1, battery B, and meter M_1 must be reversed. Also, the

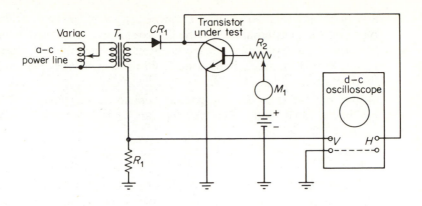

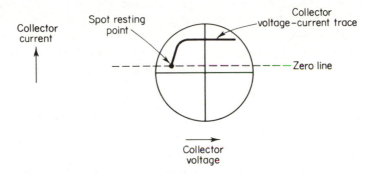

FIGURE 9-1 Testing transistors for collector current versus emitter current.

horizontal zero reference point should be at the right of the screen rather than at the left.

The following procedure displays a single curve. There are commercial transistor test units, or *curve tracers*, that trace several curves (one for each value of emitter current). Such curve tracers are described throughout this chapter. Here we concentrate on the use of a conventional oscilloscope to perform a similar curve-tracing function.

1. Connect the equipment as shown in Fig. 9–1.
2. Place the oscilloscope in operation (Chapter 5). Voltage-calibrate both the vertical and horizontal channels as necessary (Chapter 6). The spot should be at the vertical center, and at the left (for NPN) of the horizontal center with no signal applied to either channel.

3. Switch off the internal recurrent sweep. Set the sweep selector and sync selector to external. Leave the horizontal and vertical gain controls set at the (voltage) calibrate position. Set the vertical polarity switch so that the trace deflects up from the centerline as shown in Fig. 9–1a.

4. Adjust the variac so that the voltage applied to the collector is the maximum rated value. This voltage can be read on the voltage-calibrated horizontal scale.

5. Adjust resistor R_2 for the desired emitter current as indicated on meter M_1.

6. Check the oscilloscope pattern against the transistor specifications. Compare the current–voltage values against the specified values. For example, assume that a collector current of 300 mA should flow with 7 V applied. This can be checked by measuring along the horizontal scale to the 7-V point, then measuring from that point up to the trace. The 7-V (horizontal) point should intersect the trace at the 300-mA vertical point (at the 3-V vertical point if R_1 is 10 Ω).

7. If desired, adjust resistor R_2 for another emitter current value as indicated on meter M_1. Then check the new collector current–voltage curve.

9–2. TWO-JUNCTION TRANSISTOR SWITCHING TESTS

Transistors to be used in pulse or digital applications must be tested for switching characteristics. For example, when a pulse is applied to the input of a transistor, there is a measurable time delay before the pulse starts to appear at the output. Similarly, after the pulse is removed, there is additional time delay before the transistor output returns to its normal level. These "switching times" or "turn-on" and "turn-off" times are usually of the order of a few microseconds (or possibly nanoseconds) for high-speed pulse transistors.

The switching characteristics of transistors for computer or digital work are listed on the datasheets. Each manufacturer lists its own set of specifications. However, there are four terms (rise time, fall time, delay time, and storage time) common to most datasheets for transistors used in pulse work. These switching characteristics are of particular importance where the pulse durations are short. For example, assume that the turn-on time of a transistor is 10 ns and that a 5-ns pulse is applied to the transistor input. There will be no pulse output, or the pulse will be drastically distorted.

9–2–1. Pulse and Square-Wave Definitions

The terms commonly used in describing transistor switching characteristics are illustrated in Fig. 12–1. The terms of Fig. 12–1 are defined in Sec. 12–1. The input pulse represents an ideal input waveform. The other waveforms in Fig. 12–1 represent typical output waveforms in order to show the relationship.

9-2-2.　Testing Two-Junction Transistors for Switching Time

Figure 9–2 shows two circuits for testing the switching characteristics of transistors used in pulse or computer work. Figure 9–2a is used when a dual-trace oscilloscope is available. Figure 9–2b requires a conventional, single-trace oscilloscope. In either case, the oscilloscope must have a wide frequency response and good transient characteristics.

In either circuit of Fig. 9–2, the oscilloscope vertical channel is voltage-calibrated in the normal manner, while the horizontal channel must be time-calibrated (rather than sweep-frequency-calibrated). The transistor is tested by applying a pulse to the base of the transistor under test, with a specific bias applied to the transistor base. The same pulse is applied to one of the oscilloscope vertical inputs (Fig. 9–2a) or to the pulse monitor oscilloscope (Fig. 9–2b). The transistor collector output (inverted 180° by the common-emitter circuit) is applied to the other oscilloscope vertical input (Fig. 9–2a) or to the output monitor oscilloscope (Fig. 9–2b). In the Fig. 9–2b circuit, the same oscilloscope can be moved between the two monitor points. The two pulses (input and output) are then compared as to rise time, fall time, delay time, storage time, and so on. The transistor output pulse characteristics can then be compared with transistor specifications.

9-2-3.　Guidelines for Switching Tests

Since rise-time and fall-time measurements are of special importance in all switching tests, the relationship between oscilloscope rise time and the rise time of the transistor must be taken into account. This subject is discussed thoroughly in Sec. 12–1.

9-3.　TWO-JUNCTION TRANSISTOR TESTS USING CURVE TRACERS

As discussed in Chapter 1, the most practical means of measuring transistor characteristics with an oscilloscope is to display the characteristics as *curve traces*. Since the oscilloscope screen can be calibrated in voltage and current, the transistor characteristics can be read from the screen directly. If a number of curves are made with an oscilloscope, they can be compared with the curves drawn on transistor datasheets.

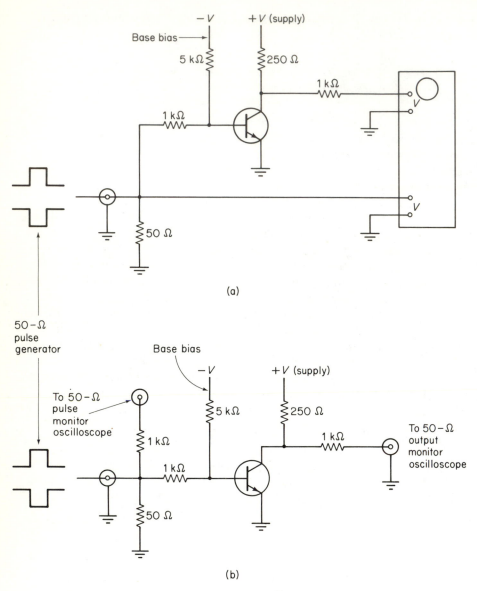

FIGURE 9-2 Testing two-junction transistors for switching characteristics: (a) dual-trace; (b) single-trace.

9-3-1. Basic Curve Tracer Operating Procedure

The following is a typical or basic operating procedure for a curve
tracer. More detailed procedures for test of specific transistor characteristics
using a curve tracer are given in the remaining sections of this chapter. The

160

use of curve tracers to test characteristics of other devices (FETs, UJTs, SCRs, etc.) are also given in this chapter.

1. As shown in Fig. 9–3, the transistor is connected into a grounded-emitter circuit. Precise 1-mA current steps are applied to the base. Voltage sweeps from 0 to approximately 5 V are applied to the collector. The oscilloscope vertical deflection is obtained from the

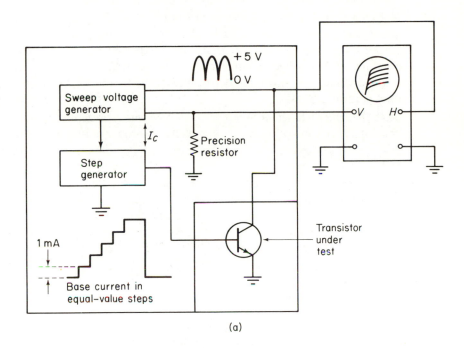

(a)

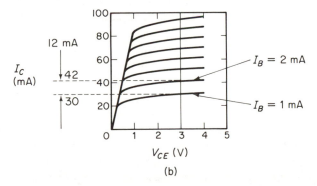

(b)

FIGURE 9–3 (a) Basic test connections and (b) typical oscilloscope displays for transistor curve tracer.

resultant collector current. All of this is done by setting of switches on the curve tracer.

2. On some curve tracers, each sweep must be initiated individually, whereas other instruments produce a series of 10 (or more) curves in sequence automatically.

3. Once the curves are made, they are interpreted as follows.

4. Choose the vertical line corresponding to the specified collector voltage. For example, the 3-V vertical line is chosen in Fig. 9–3.

5. Note (on that line) the distance between the two curves that appear above and below the specified base current (or specified collector current). For example, assume that the transistor of Fig. 9–3 is to be operated with a collector current of 30 to 40 mA, with 3 V at the collector. The two most logical curves are then the 1- and 2-mA base current curves.

6. Note that the distance between these two curves represents a difference of 12 mA in collector current. (The 1-mA curve intersects the 3-V line at 30 mA, and the 2-mA curve intersects the 3-V line at 42 mA.)

7. Divide the collector-current difference by the base-current difference. Since the base current per step is 1 mA, and the difference in collector current is 12 mA, the gain (beta) is 12. If the base current per step is 0.1 mA (as it is for some curve tracers), and all other conditions are the same, the beta is 120 (12/0.1 = 120).

9–3–2.   *NPN versus PNP Transistors*

On a typical curve tracer, the curves of an NPN transistor are in a positive direction. That is, zero volts is at the left and zero current is at the bottom of the display. The curves sweep to the right and upward as collector voltage and current increases from zero. The collector sweep voltage is of positive polarity.

The family of curves of a PNP transistor is in the negative direction. That is, zero volts is at the right and zero current is at the top of the display. The curves sweep to the left and downward as collector voltage and current increase from zero. The collector sweep voltage is of negative polarity.

All transistor tests described in this chapter apply equally to NPN and PNP transistors. Any examples showing only an NPN or PNP type should be understood to apply to the counterpart as well. Basic characteristics of both types are the same, although displays are inverted with reference to each other. Therefore, any test that can be made for NPN transistors can also be made for PNP transistors, and vice versa.

9-3-3.   Becoming Familiar with the Curve Tracer

The best way to become familiar with the curve tracer is to gather an assortment of known good transistors (as well as FETs, UJTs, SCRs, etc.) and test them using the curve tracer. Carefully observe the normal test results, the effects of readjusting the controls, any peculiarities of various semiconductors, and so on.

Generally, transistors and other semiconductor devices are tested with respect to manufacturers' specification sheets, which state certain conditions of the test and minimum performance standards. Keep in mind that it is normal for characteristics of some devices to vary quite widely from one semiconductor to another even though they have the same type number. Also, manufacturers' specification sheets are not always readily available, and the manufacturer and type number are not always easily determined. However, the semiconductors may be tested and determined to be good or bad, even when the type number and specifications are not known. This is because most transistor failures are of the catastrophic type (shorts, opens, or leakage). The curve tracer shows such failures immediately, as described in other paragraphs of this chapter. A curve tracer can also identify a good transistor as NPN or PNP, and can show the normal operating current range.

9-3-4.   Typical Transistor Test Procedure

The following describes a typical operating procedure for a curve tracer when used to test a transistor. Although the procedures are for the B&K Precision curve tracer, they are typical for a great variety of curve tracers.

1. Connect the equipment as shown in Fig. 9–4. Turn on the curve tracer and oscilloscope. Calibrate the oscilloscope if not already calibrated.

2. Set the curve tracer vertical sensitivity and step selector controls to the "fast-setup" markers (as described in Chapter 5). Set the sweep voltage control to zero.

3. Plug the transistor into the left or right socket on the curve tracer, or connect test leads from the left or right C, B, and E jacks to the collector, base, and emitter of the transistor. Set the socket switch to select the socket or jacks in use.

4. Set the polarity switch to NPN or PNP to match the transistor type. Slowly increase the sweep voltage control until a complete display appears on the oscilloscope screen. If the type of transistor is not known, do not exceed the fast-setup marker of the sweep voltage control (20 V). If a display does not appear, return the sweep

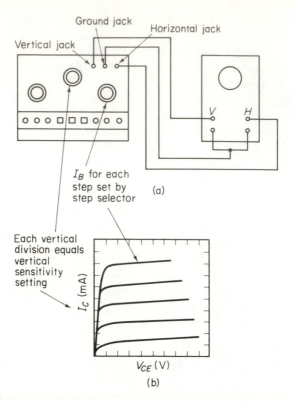

Ground jack

Horizontal jack

Vertical jack

I_B for each
step set by
step selector

(a)

Each vertical
division equals
vertical
sensitivity
setting

I_C (mA)

V_{CE} (V)

(b)

FIGURE 9-4 Typical (a) test connections and (b) display for curve
tracer test of NPN transistor.

voltage to zero, reverse the polarity, and slowly increase the sweep
voltage again until a complete display appears.

5. Adjust the centering controls of the oscilloscope to position the
    family of curves properly, as shown in Fig. 9–5. For NPN tran-
    sistors, the display should be positioned so that the curves start at
    the left-hand bottom corner of the graticule scale. For PNP tran-
    sistors, the display should be positioned so that the curves start at
    the right-hand top corner of the graticule scale.

6. If there is only a single vertical trace as shown in Fig. 9–6, the tran-
    sistor is probably shorted. With a shorted transistor, the current
    (vertical trace) increases sharply to the current limiting value deter-
    mined by the curve tracer. There is no family of curves, since the
    value of base current applied has no effect.

7. If there is only a single horizontal trace as shown in Fig. 9–7, the
    transistor is probably open. With an open transistor there is no
    collector-current flow, and thus no vertical displacement of the
    display.

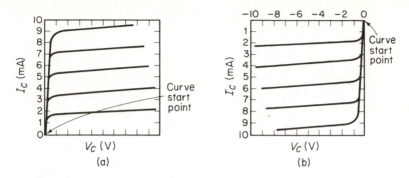

FIGURE 9-5 Typical characteristic curves for NPN and PNP transistors: (a) NPN; (b) PNP.

9-3-5. Effects of Curve Tracer Controls

The following is a summary of the effects of curve tracer controls on the display when testing two-junction transistors.

If the *polarity* switch is set to the wrong polarity, the oscilloscope displays reverse breakdown characteristics of the transistor base–emitter junction. For familiarization, check several transistors, placing the polarity switch in both the correct and incorrect positions and observing the results.

The *sweep voltage* control adjusts the peak collector voltage (V_c), and thus the horizontal width of the display. Increasing the setting widens the display and may cause the display to go off-scale. Decrease and increase the setting of the control and note the effect upon the display. If increasing the setting drives the display off-scale, recalibrate the oscilloscope horizontal axis to a higher voltage. As the sweep voltage is increased, collector

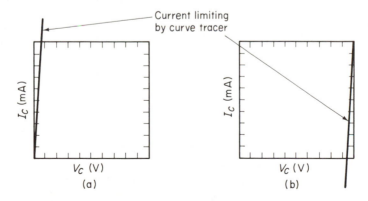

FIGURE 9-6 Typical curve tracer display for a shorted transistor: (a) NPN; (b) PNP.

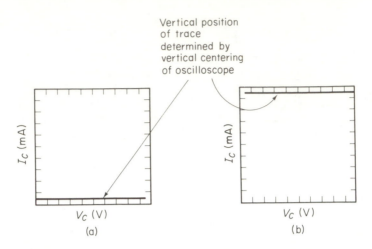

Vertical position
of trace
determined by
vertical centering
of oscilloscope

I_C (mA)

V_C (V)

(a)

I_C (mA)

V_C (V)

(b)

FIGURE 9-7 Typical curve tracer display for an open transistor:
(a) NPN; (b) PNP.

breakdown will be observed on the oscilloscope display, as discussed in Sec. 9-5. Collector breakdown is observed as a sharp turn in the curves (sharp decrease in collector current). Most tests are conducted below collector breakdown voltage. After observing the effects of collector breakdown, reduce the setting somewhat below the breakdown point.

The setting of the *vertical sensitivity* control is determined primarily by the physical size of the transistor or semiconductor being tested. Most small-signal transistors, including most small plastic-case transistors, should be tested at the 1-mA/div position. Some small-signal transistors, including several metallic case transistors, should be tested at the 2-mA/div position. The 5-mA/div and 10-mA/div positions are primarily for power transistors. The 5-mA/div position is for low- to medium-power transistors, and those with self-contained heat-radiating cases. High-power transistors should be tested at 10 mA/div.

Generally, power transistors may be tested without their usual heat sinks if testing is limited to a few seconds, just long enough to make the reading. It is especially important to keep testing time short during periods of voltage breakdown and current limiting, to prevent overheading and thermal runaway damage. The B&K Precision 501A curve tracer uses current limiting in the collector voltage supply to prevent damage to the transistor being tested. However, the point of current limiting is increased with each step of the vertical sensitivity control. Thus, the lowest acceptable setting of the control must be used to protect small transistors. Start with the lowest position (1 mA/div) and increase to a higher setting only as necessary.

The *step selector* switch selects the base current steps. The approximate setting of the step selector is also somewhat related to the physical size of the transistor. The fast-setup position (10 μA per step) is a good starting point for most signal transistors. Very high gain, small-signal transistors require a lower setting. Large-signal and power transistors require a higher setting.

If transistor specification sheets are available, it is desirable to use the manufacturer's data, and set the step selector to produce the specified collector current where beta is to be measured (Sec. 9-4). If data sheets are not available, the general rule is to adjust the step selector for the most curves displayed on the vertical sensitivity range being used. If the setting is too high, some of the curves may reach the current-limiting value and be superimposed on each other, causing fewer than five curves to be displayed. When more than one position displays all five curves, select the position that produces the most *even spacing between curves.*

The *volts per step* positions of the step selector are for testing FETs (Sec. 9-11). In the volts per step positions, step polarity is reversed with respect to the sweep voltage. The method of adjustment is the same. Set for the maximum number of curves and the most even spacing between curves on the vertical range being tested.

9-3-6. *Sorting and Matching Transistors with a Curve Tracer*

A special method of setting up the oscilloscope can be used when transistors are to be matched with a curve tracer. This technique is especially helpful when trying to match the gain of two opposite polarity devices (NPN versus PNP). The basic setup procedure is as follows:

1. Calibrate the oscilloscope vertical axis for 1 V full scale, and the horizontal axis for 30 V full scale.
2. Adjust the oscilloscope positioning controls to place the CRT trace (dot, with no inputs) at the screen center.
3. Adjust the curve tracer sweep voltage control until the trace sweeps out to either edge of the graticule, depending on the position of the polarity switch.
4. Set the vertical sensitivity control to the 2-mA/div position.
5. Set the step selector to the 10-μA position.

As shown in Fig. 9-8, when the special oscilloscope setup is used, NPN transistors display curves in the upper right-hand quarter of the graticule, and PNP in the lower left-hand quarter. The polarity of an unknown device can be determined immediately by inserting the device into either socket, switching to that socket, and moving the polarity switch back and forth until curves appear.

Once the curves are displayed as shown in Fig. 9-8, the step selector may need readjustment to bring all six curves into view on the screen. Note that when the polarity switch is set to PNP for an NPN device (and vice versa) a

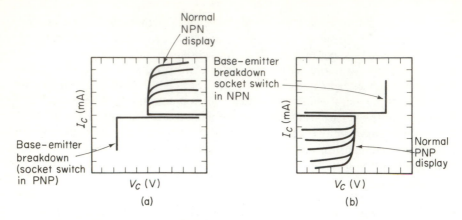

FIGURE 9-8 Typical displays for sorting and matching transistors with a curve tracer: (a) NPN; (b) PNP.

single curve appears on the screen opposite to where normal curves appear, as shown in Fig. 9–8. This represents the base–emitter breakdown voltage of the device under test.

Matching two complementary devices may be done by inserting the PNP unit in the right socket and the NPN unit in the left socket. Simultaneously switch both the polarity and socket switches in unison while watching the display. When the devices are matched, both sets of curves appear to be the same but opposite in polarity, as shown in Fig. 9–9. The external transistor jacks can also be used (instead of the sockets), if desired.

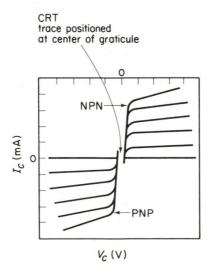

FIGURE 9-9 Matching NPN and PNP transistors.

9-3-7. In-Circuit Transistor Testing with a Curve Tracer

It is possible to use a curve tracer to test a transistor in-circuit on a go–no go basis. Curves obtained with a transistor in-circuit most often appear badly distorted due to in-circuit impedances (capacitors, resistors, inductors, etc.). However, in-circuit curves can be used to show transistor action or inaction. Some transistor circuits will not produce any readable curves. For example, a transistor used as a series pass regulator in a power supply will appear shorted because the large-value filter capacitors across the transistor act as a low impedance to the curve tracer sweep signal. Another example is where the curve tracer step drive signal is shunted away by a very low in-circuit base–emitter resistance. In these cases, it is best to remove the transistor from the circuit for test on the curve tracer.

When making in-circuit tests, it may be necessary to readjust the sweep selector to produce curves. However, the sweep voltage should not require readjustment for in-circuit tests. Also, it is not recommended that breakdown voltage tests be performed with the device in-circuit.

Figure 9–10 shows some typical in-circuit curves of normally functioning transistors. The loops in the NPN curves are caused by capacitance in the transistor collector circuit. Note that five loops and a baseline can be seen in the NPN display. This indicates that each base step produces some collector current, meaning that the transistor is probably good. The PNP curves, typical of those for a transistor oscillator, indicate severe leakage, even though the transistor is not leaking.

The best way to interpret in-circuit curve displays is through experience. Practice with several types of transistors and circuits, keeping a list of test conditions (sweep voltage, current per step, circuit adjustment settings, etc.) for each circuit of the equipment being tested.

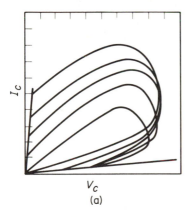
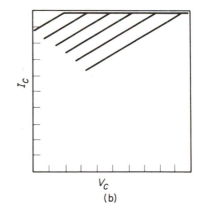

I_C　　V_C
(a)

I_C　　V_C
(b)

FIGURE 9-10 Testing transistors in-circuit with a curve tracer. (a) NPN: loops caused by capacitance in the collector circuit; (b) PNP: transistor oscillator circuit with leakage.

9-4. TRANSISTOR CURRENT-GAIN TEST USING A CURVE TRACER

The current gain of a transistor is the single most important characteristic and is usually measured before any other tests are performed. The general condition of a transistor can most often be determined while testing for current gain. There are two generally accepted ways to measure current gain: *d-c beta* or direct-current gain, and *a-c beta* or alternating-current gain. The following paragraphs describe both test procedures using a curve tracer.

9-4-1. D-c Beta Tests

D-c beta can be defined as the ratio of collector current to base current measured at *one specific point* of collector voltage and current. The validity of d-c beta depends on the point of measurement, and this point is usually specified on manufacturer's datasheets. The measurement point usually centers about the typical operating range for which the transistor is designed. For example, a typical d-c beta is specified as h_{FE} = 40 (min.) 125 (typical) and 400 (max.) with an I_C of 5 mA and V_C of 10 V.

Figure 9–11 shows the technique of d-c beta measurement using a curve tracer. The point of measurement is conveniently centered on the display by using 1 mA/division vertical and 2 V/division horizontal calibration. With each step at 10 μA, the third curve (30 μA) passes through the measurement point. Thus, 30 μA of base current produces 5 mA of collector current at the 10-V collector point. Using these values, d-c beta = I_C/I_B = 5 mA/0.03 mA = 166.

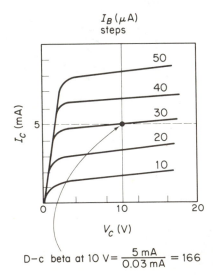

D-c beta at 10 V = $\dfrac{5\,\text{mA}}{0.03\,\text{mA}}$ = 166

FIGURE 9-11 D-c beta measurement using a curve tracer.

9-4-2. A-c Beta Tests

The a-c beta or dynamic current gain of a transistor can be defined as the *ratio of change* in collector current to the *change* in base current at a specified collector voltage. A-c beta is generally more useful than d-c beta since the transistor is tested under actual operating conditions. This provides a better basis for predicting transistor performance. Figure 9–12 shows the technique of a-c beta measurement using a curve tracer.

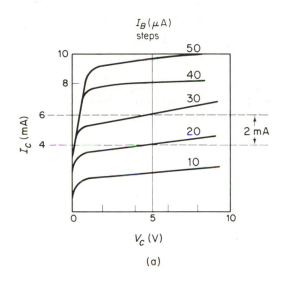

(a)

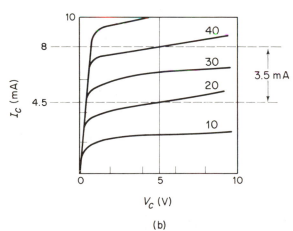

(b)

FIGURE 9-12 A-c beta measurement using a curve tracer. (a) A-c beta at 5 V = 2/0.01 = 200; (b) a-c beta at 5 V = 3.5/0.02 = 175.

1. Measure the difference in collector current (ΔI_C) between the two curves on the oscilloscope display. (The settings of the curve tracer and/or oscilloscope controls determines the amount of collector current represented by each vertical division of the display scale. In Fig. 9–12, each vertical division represents 2 mA of collector current.) Be sure that *both curve readings* are taken at the same collector voltage. (In Fig. 9–12, both readings are taken at 5 V.)

2. Note the change in base current (ΔI_B) that produces each curve. In Fig. 9–12a, each curve is produced by a change of 10 μA. Thus, ΔI_B is 10 μA for any of the curves in Fig. 9–12a.

3. Calculate beta by dividing ΔI_C by ΔI_B. For example, if ΔI_C is 2 mA and ΔI_B is 10 μA, as shown in Fig. 9–12a, beta is 200 (2/0.10 = 200).

It is generally easier to use the two centermost curves of the display to measure a-c beta. However, this is not always practical. If so, the ΔI_C measurement may be made between two nonadjacent curves. For example, the difference between the collector current of the second and fourth curves may be used for measurement of ΔI_C, as shown in Fig. 9–12b. If this method is used, make certain to use *two steps* of base current for determining ΔI_B when calculating beta. Using the values of Fig. 9–12b, a-c beta = 3.5 mA/20 μA = 175 (at V_C of 5 V).

If the transistor datasheet is available, measure the a-c beta at the approximate collector current and voltage specified. If no values are specified, adjust the curve tracer controls for a display of the most evenly and widely spaced curves.

9-4-3. Inconsistent A-c Beta

Note that the beta (a-c or d-c) of any transistor is not constant. Beta always depends upon the point of measurement. The distance between all curves is seldom equal, which means that ΔI_C is not the same at various regions of base current and, to a lesser extent, at various collector voltages. When the curves are closer together, the transistor is operating in a region of lower gain. Generally, gain is the highest at the normal operating region of a transistor, and is lower at points above and below the normal operating region.

9-4-4. Testing Transistor Linearity and Distortion

The curves produced during a-c beta tests can be used to measure transistor linearity and distortion. As discussed, beta is not necessarily constant, but may vary with changes in voltage and current. When such variation of gain occurs, the transistor is nonlinear and introduces distortion if operated in the nonlinear region. To measure linearity and possible distortion, adjust the curve tracer controls for the most evenly spaced display of curves, or so that the centermost curves of the display are the most evenly spaced, as shown in Fig. 9–13. Note that it is not always possible to obtain evenly spaced curves on any part of the display for some transistor types. Set the sweep voltage to approach, but not reach, collector breakdown (Sec. 9–5).

Note that the curves of Fig. 9–13 are somewhat shorter at higher collector currents. Plot an imaginary line along the ends of the curves as shown in Fig. 9–13. This line is called the test load line. Plot an operating load line in parallel with the test load line, but intersecting the zero I_C line at the rated operating voltage (at about 13 V in Fig. 9–13). Pick three successive curves that represent the normal operating base-current region for the transistor (probably those curves with the most even spacing—10, 20, and 30 μA in Fig. 9–13). Measure and compare the change in collector current (ΔI_C) between the curves along the operating load line.

As an example, first measure between the 10- and 20-μA curves, then between the 20- and 30-μA curves. If the I_C values are equal, the transistor is operating in a linear region, and is free of distortion. If the I_C values are not equal, distortion will be introduced. The greater the difference in ΔI_C values, the greater the distortion.

Measuring Percentage of Distortion. It is possible to measure the approximate percentage of distortion using a family of curves such as shown in Fig. 9–13. The theoretical input and output sinewaves superimposed on the

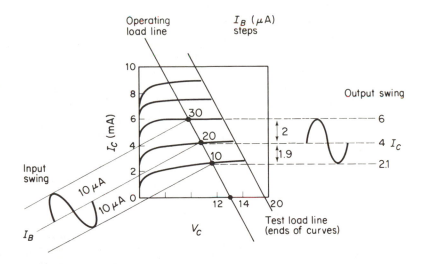

FIGURE 9–13 Measuring transistor linearity and distortion using a curve tracer.

curves assume that the transistor is operating in a basic common-emitter amplifier circuit, that the static or no-signal base current I_B is 20 μA, and that the no-signal collector current I_C is 4 mA. A theoretical sine-wave signal of 20 μA peak to peak is applied to the base, causing the base current to swing from 10 μA to 30 μA. (In reality, the swing is caused by the 10-μA current steps applied to the base.)

Assuming a beta of 200 and no distortion, the output collector current should then swing from 2 mA to 6 mA, or 2 mA above and below the no-signal point of 4 mA. However, as shown in Fig. 9–13, the output collector current swings from 2.1 mA to 6 mA, producing a total swing of 3.9 mA. Thus, there is a 0.1-mA imbalance (the output negative swing is 1.9 mA instead of 2 mA). This amounts to approximately 2.5% distortion (0.1/3.9 = 0.0256). Such distortion might be quite acceptable for some applications (class C amplifiers, frequency multipliers, switching transistors, etc.) but not for other applications (class A and B audio amplifiers).

Using the Operating Load Line.   The advantage of using the operating load line rather than some specific value of collector voltage is that the operating load line nearly duplicates the dynamic conditions of operation. The transistor operates with a fixed load and with a fixed supply voltage. However, with a signal present, the collector voltage does not remain fixed. An increase in collector current reduces collector voltage, and vice versa. However, the load line follows the variation in collector voltage produced by the changing signal.

If a particular transistor shows nearly horizontal collector-current curves, there is little difference between using the operating load line or some specific collector voltage. However, in transistors with more slope to the collector-current curves (which is usually the case), the collector voltage affects the collector current, and can produce considerably different results in making linearity and distortion measurements.

9–5. TRANSISTOR BREAKDOWN VOLTAGE TESTS USING A CURVE TRACER

The variable collector sweep voltage of a curve tracer can be used to test for breakdown voltage. As sweep voltage is increased, a collector breakdown will be reached. Of course, the value at which this occurs depends upon the transistor type. A typical curve tracer, such as the B&K Precision 501A, tests breakdown up to 100 V, which is sufficient to test all but the high-voltage-rated transistors. At the collector breakdown voltage, the collector current becomes independent of base current, and rises sharply to the current-limiting protection limit of the curve tracer. (If the curve tracer is not provided with a current-limiting feature, the transistor can be destroyed if the collector voltage is maintained at or above the breakdown point for any length of time.)

Figure 9-14 shows a typical family of curves with the sweep voltage set high enough to cause collector breakdown. In the examples shown, breakdown occurs at a collector voltage of approximately 35 to 40 V for the transistor with abrupt breakdown. Note that base current has little effect upon the point at which the increase in collector current occurs, particularly with transistors having an abrupt breakdown characteristic. As shown in Fig. 9-14, it is sometimes difficult to determine the precise breakdown voltage for transistors with a gradual breakdown characteristic.

To perform the measurement, first adjust the curve tracer controls for a normal family of curves on the display. For best results, the normal display should not fill the scale horizontally. Next, increase the sweep voltage control until the upturn in collector current (at the tail of the curves) is observed. As shown in Fig. 9-14, this upturn is very sharp for most transistors, but can be gradual for a few transistor types.

Read the collector voltage value at which the upturn occurs, using the horizontal scale. If a breakdown voltage specification for the transistor is available, use Fig. 9-14 to determine if the transistor is acceptable for a particular application. If no specification is available, a simple guideline is that the transistor should withstand approximately twice the collector supply voltage of the circuit in which the transistor is to be used.

When making any breakdown test, *keep the test as short as possible.* Even with current limiting found in most curve tracers, the current value is much higher than normal and causes a temperature increase in the transistor. This can result in damage to the transistor.

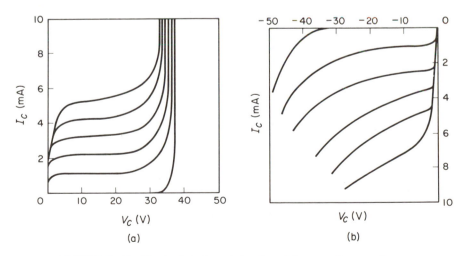

FIGURE 9-14 Measuring transistor breakdown voltage using a curve tracer: (a) abrupt breakdown; (b) gradual breakdown.

9-6. TRANSISTOR LEAKAGE TESTS USING A CURVE TRACER

The zero-base-current curve of a curve tracer can be used to test the collector leakage of a transistor. Collector leakage current is the collector-to-emitter current that flows when the transistor is supposed to be completely off, and is listed as I_{CEO} on most datasheets. If the transistor is leaking, increasing collector voltage causes the collector current to increase independently of the base current. Some collector leakage is normal for most germanium transistors, but not for silicon transistors. Excessive leakage in silicon transistors is usually a sign of some defect. When measured in relation to datasheet specifications, all leakage tests should be made at the specified collector voltage and temperature. Leakage current generally increases with increases in temperature.

Although any sloping line or curve can be used for measurement, leakage can be measured best by observing the zero-base-current line as shown in Fig. 9–15. This is because there should be no change in collector current (the curve should remain horizontal) when the collector voltage is increased, if there is zero base current and no leakage. Thus, any slope of the zero-base-current curve indicates some leakage. In the example of Fig. 9–15, there is approximately 0.5 mA leakage at 10 V and about 0.25 mA leakage at 5 V.

When making a transistor leakage test using a curve tracer, it is possible that some leakage may be indicated, even though the transistor is not leaking. This is because the horizontal input impedance or resistance of the oscilloscope is connected across the collector–emitter of the transistor, as shown in Fig. 9–3. To eliminate any doubt, switch the transistor in and out of the cir-

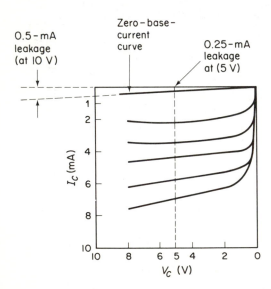

FIGURE 9–15 Measuring transistor leakage using a curve tracer.

cuit while observing the zero-base-current curve. If the curve remains the same with the transistor switched in and out, the indicated leakage is caused by the oscilloscope impedance.

9-7. *TRANSISTOR SATURATION-VOLTAGE TESTS USING A CURVE TRACER*

A family of curves produced on a curve tracer can be used to test saturation voltage of a transistor. The collector saturation region of a transistor is that portion of the family of curves in the area of low collector voltage and current below the knee of each curve, and is listed as $V_{CE(sat)}$ on most datasheets. This is shown in Fig. 9-16. Notice that the knee of each curve occurs at approximately the same collector voltage, regardless of base current. Also, collector voltage above the knee has little effect upon collector current. Instead, base current has the predominant effect.

Saturation voltage is the collector voltage at the knee of the curve. For measurement in comparison to specifications, both base current and collector current should be stated. The specification value is the maximum value at which the knee should occur. Thus, if the specification value is on or above the knee, the transistor is acceptable.

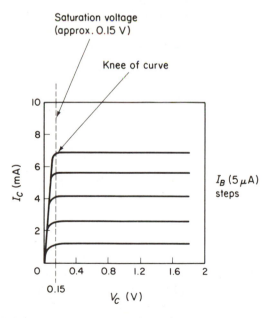

FIGURE 9-16 Measuring transistor saturation voltage using a curve tracer.

To measure saturation voltage on a curve tracer, only the saturation region need be displayed. This is the low-collector-voltage portion, up to and including the knee of each curve. If curve tracer controls permit, the display should be spread out using a low-voltage horizontal calibration (collector voltage) such as 0.2 V/division, as shown in Fig. 9–16. This provides the most accurate means of measuring the low collector-voltage value that produces saturation.

If desired, you can calculate the saturation resistance, listed as $r_{CE(sat)}$ on some datasheets. Saturation resistance equals the collector voltage divided by the collector current for a given value of base current in the collector satura-collector current (I_C) of 4 mA, the saturation resistance is 37.5 Ω (0.15/01004 = 37.5).

Generally, saturation resistance is stated as a maximum acceptable limit. *Dynamic resistance* is found by calculating or plotting the average saturation resistance over a range of base current.

9-8 TRANSISTOR OUTPUT ADMITTANCE AND IMPEDANCE TESTS USING A CURVE TRACER

Output admittance of a transistor can be measured from the same family of curves as displayed for gain or beta measurement. The dynamic output admittance of a transistor, listed as h_{oe} on some datasheets, is the measurement of the change in collector current (ΔI_C) resulting from a specific change in collector voltage (ΔV_C) as a constant base current, as shown in Fig. 9–17. Admittance is measured in mhos. Using the values of Fig. 9–17, the output admittance or h_{oe} is 750 $\mu\Omega$ when the base current is a constant 150 μA.

Output impedance of a transistor is the reciprocal of the output admittance, and is measured in ohms. Output impedance can be calculated by transposing the current and voltage values used in determining output admittance. Using the values of Fig. 9–17, the output impedance is 1.333 kΩ (4/0.003 = 1333). Note that a transistor provides maximum power transformation when the load impedance equals the output impedance.

A change in collector voltage normally causes change in collector current. For some transistors, the effect is quite apparent because the curves have a noticeable slope. Such transistors have a comparatively high output admittance. Other transistors show a nearly horizontal curve with a very small change in collector current. Such transistors have a low output admittance.

Both output impedance and admittance measurements are taken at a constant base current (along one of the display curves). If datasheet information is used for reference, use the base current specified. Otherwise, select a base-current curve that is typical for the normal operating range of the transistor being tested. Make the measurement between two specific collector voltages (between 3 and 7 V on the 150-μA curve in Fig. 9–17). When testing per datasheet specifications, use the specified voltages. Without specifica-

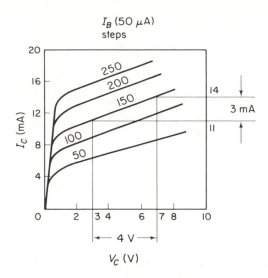

FIGURE 9-17 Measuring transistor output admittance and impedance using a curve tracer. Output admittance = 3 mA/4 V = 0.003/4 = 0.00075 = 750 $\mu\Omega$; output impedance = 4 V/3 mA = 4/0.003 = 1333 Ω.

tions, select two voltages, one just above the saturation knee of the curve and one somewhat below collector breakdown.

If you must measure very small collector-current values (I_C), it may be necessary to use a higher vertical gain for the oscilloscope. However, the oscilloscope vertical must remain calibrated.

9-9 TESTING EFFECTS OF TEMPERATURE ON TRANSISTORS USING A CURVE TRACER

When current is conducted by a transistor, some heat is generated. The amount of heat increases with the value of collector voltage and current. An excessive heat buildup can result if the transistor cannot dissipate the heat generated. If excessive heat is generated while testing with a curve tracer, the results are easily detected in the display.

A high temperature can produce a noticeable loop in the curves as shown in Fig. 9–18a. Collector capacitance or inductance can also cause a loop in the curves but can be distinguished from a temperature loop. *For a temperature loop,* the loop size decreases and disappears when the base current steps or collector voltage are reduced. The reason for the loop in the curves is that collector current does not increase and decrease at the same rate when the sweep voltage is applied.

Under normal conditions, the transistor is cool when sweep voltage starts from zero. As the sweep voltage increases to maximum, a collector cur-

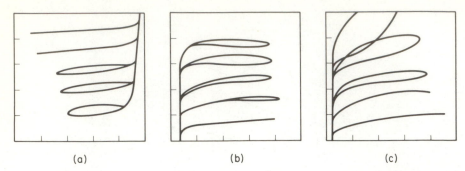

FIGURE 9-18 Testing effects of temperature on transistors using a curve tracer. (a) High-temperature loops; (b) high-temperature droop; (c) thermal runaway (vertical roll effect).

rent sweep is made. Meanwhile, the temperature is increasing. The temperature increase causes an additional amount of collector current. As the sweep voltage returns to zero, the collector current decreases, but with a lag. During the return sweep, the temperature drops from maximum to normal, but a time lag is required for cooling to occur. Thus, the top portion of the loop is the increasing sweep current, and the bottom portion is the decreasing sweep current.

Another effect in some transistors is that collector current droops at the high end of the curve. In this case, the increase in temperature causes a decrease in collector current, as shown in Fig. 9–18b.

Excessive current to the extent that *thermal runaway* begins is observed as a "vertical roll" effect, as shown in Fig. 9–18c. Thermal runaway results when an increase in temperature, caused by increased current, results in further current increases (and a further temperature increase). If thermal runaway continues, the transistor will be burned out or permanently damaged. When thermal runaway starts as a transistor is being tested on a curve tracer, the entire family of curves moves in the direction of higher collector current (as temperature buildup continues).

CAUTION

Turn off the curve tracer immediately if you observe any signs of a thermal runaway (a vertical roll of all curves, usually starting with roll of the highest base-current curves).

9-10. FET SWITCHING-TIME TESTS

Switching-time characteristics are of particular importance for FETs used in digital electronics. The datasheets that describe FETs specifically designed for switching include *timing characteristics*. Typically, these include t_d (delay

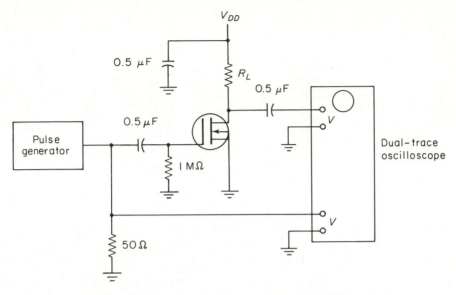

FIGURE 9-19 Test circuit for measurement of FET switching characteristics. R_L = value that causes negligible d-c drop at I_{DSS}.

time), t_S (storage time), t_r (rise time), and t_f (fall time). These characteristics are measured with a pulse source and a multiple-trace oscilloscope, as are similar characteristics for two-junction transistors (Sec. 9-2). A typical test circuit for measurement of FET switching characteristics is shown in Fig. 9-19.

9-11. FET TESTS USING A CURVE TRACER

In many respects, the testing of FETs using a curve tracer is similar to testing two-junction transistors. A family of curves is displayed on the oscilloscope, and the curves have a similar appearance. As shown in Fig. 9-20, N-channel FETs have a family of curves similar to NPN transistors, and P-channel FETs are similar to PNP transistors. Two-junction transistor curves are a graph of collector current versus collector voltage, at various base currents. FET curves are a graph of drain current versus drain voltage at various gate voltages. Similarly, FET breakdown voltage may be observed and measured by the same method used for transistors.

In several other respects, testing FETs is different from testing two-junction transistors. For FETs, the *step selector* switch (or whatever the curve tracer control may be called) is placed in a VOLTS PER STEP position. That is, the curve tracer supplies *constant-voltage* steps to the FET rather than constant-current steps (as in the case of two-junction transistors). Also, the polarity of the step voltage is reversed in relation to the sweep voltage. While

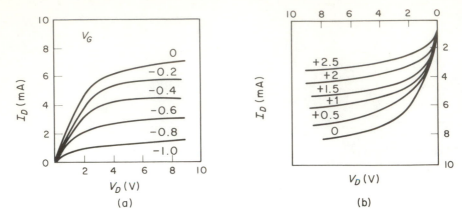

FIGURE 9-20 Typical FET characteristic curves: (a) N-channel; (b) P-channel.

the zero-base-current step of a two-junction transistor usually provides no collector current, the zero-volts step at the gate of most FETs produces the highest drain current. Each reverse bias voltage step results in less drain current, and when the gate voltage is sufficiently high, drain current is pinched off. The point of *pinch-off* (sometimes listed as V_P on FET datasheets) can be measured with the curve tracer.

The method of gain measurement for a FET is similar to the gain measurement of a two-junction transistor, but the forward transconductance of a FET has a voltage input characteristic which cannot be directly compared with the beta of a two-junction transistor (which has a current input characteristic).

9-11-1. Typical FET Test Procedure

The following steps apply to a curve tracer such as the B&K Precision 501A, but can be used as a guide for using other curve tracers to test FETs.

1. Before plugging in or connecting the FET to the curve tracer, set the controls as follows: sweep voltage to zero, vertical sensitivity to 1 mA/div, step selector, I_{DSS}/I_{CES}, and polarity to the type of FET being tested.

2. Plug the FET to be tested into the appropriate curve tracer socket. On some curve tracers, the FET is connected to the D, G, and S jacks with test leads. Use the FET basing diagram, if available, to identify the gate (G), drain (D), and source (S) leads. On some curve tracers, the socket pins are labeled for FETs as well as two-junction transistors. If there is a separate test lead for the FET shield, clip a

test lead to the shield pin and ground the lead to the appropriate curve tracer terminal. Do not leave the FET shield ungrounded.

3. Increase the sweep voltage setting a nominal amount to produce a horizontal display, but stay well below breakdown voltage (typically 5 or 10 V of sweep voltage is sufficient). A single curve should be displayed on the oscilloscope. If the FET type is unknown, try both the N-channel and P-channel positions of the polarity switch to obtain the curve.

4. Position the curve as required with the oscilloscope centering controls. If the curve extends off-scale vertically, move to the next position of the vertical sensitivity control.

5. Increase the sweep voltage as desired, but reduce the setting if breakdown is observed. (Breakdown for FETs is similar to that of two-junction transistors, as described in Sec. 9–5.)

6. Rotate the step selector control to the volts per step position, increasing the setting until a family of six curves is observed with the greatest attainable spacing between curves. If the setting is too low, the curves will be too closely spaced to take a reading. If set too high, some of the gate voltage steps may exceed pinch-off and result in fewer than six curves being displayed.

9–11–2. Important Considerations in Testing FETs

In general, FETs are more susceptible to damage from excess voltage or current than are two-junction transistors. Starting with the recommended control setting eliminates the possibility of applying test signals that are too high. Control settings should be increased (after the FET is inserted) only as much as is necessary to make the tests. Note that some MOSFETs can be damaged by a voltage transient from a static charge carried by the person handling the FET. To safeguard against such damage, discharge any static charge by touching ground with one hand before and while handling the MOSFET with the other hand.

The I_{DSS}/I_{CES} position of the step selector control displays the drain current of the FET with the gate shorted to the source. (This is typical for most, but not all, FET curve tracers.) This is the zero-bias condition and produces a single curve on the display which is represented by the maximum drain current normally flowing through the FET. Most FETs normally operate in the depletion mode with a reverse bias. The constant-voltage steps of a reverse-bias polarity as generated by this curve tracer drive the FET into the depletion mode, with the curves showing lower resultant drain current with each successive step.

To test the few enhancement-mode FETs, the gate lead can be disconnected from the curve tracer and connected to a d-c bias supply which will provide the forward-bias voltage. Be sure that any such bias supply reference is common to the source of the FET by connecting a test load between the bias supply and the S (source) jack of the curve tracer.

For testing dual-gate MOSFETs, one gate should be grounded or biased while testing. One gate can be plugged into the socket of the curve tracer, and a test lead can be clipped to the other gate. To ground the gate, connect the test lead to the S jack of the curve tracer. To bias the gate, connect the test lead to a d-c bias supply. Varying the bias supply voltage shows the effects of simultaneous inputs on the two gates of the FET. If a d-c bias supply is used, be sure to ground the d-c reference to the S jack.

9-11-3. FET Transconductance (Gain) Measurement

The most useful and common measurement to be made for a FET is the gain measurement. The dynamic gain, or gate-to-drain forward transconductance (g_m) in the common-source configuration, is the ratio of change in drain current to the change in gate voltage at a given drain voltage. Transductance is measured in mhos. Figure 9–21 shows how typical FET curves can be used to find gain. As shown, the change in drain current (ΔI_D) is 1.5 mA (from 7 mA to 5.5 mA), for a change in gate voltage (ΔV_G) of 0.1 V (from 0.1 V to 0.2 V), at a V_{DS} of 6 V. This indicates a gain (transconductance, or g_m) of 15,000 μmhos (0.0015 A/0.1 V = 0.015 mho = 15,000 μmhos).

As with a two-junction transistor, the gain of a FET is not constant over the entire voltage and current range. The gain is normally calculated in the typical operating range. Distortion and linearity may be determined by the same method as described for two-junction transistors. That is, if the spacing between curves is equal, the FET is linear.

9-11-4. FET Pinch-off (V_P) Voltage Measurement

An important characteristic for depletion-mode FETs is the amount of gate voltage required to turn off drain current. This value is called the pinch-off voltage characteristic and may be measured from the family of curves as shown in Fig. 9–22a.

Figure 9–22a shows the display with the step selector set at 0.5 V per step. Note that the entire family of curves is displayed, and that drain current continues to flow at the highest step of − 2.5 V. Figure 9–22b shows that when the step selector is increased to 1 V per step, the entire family of curves is not displayed. In fact, the − 3-V, − 4-V, and − 5-V curves are superimposed upon each other at zero drain current. From this it can be concluded that pinch-off occurs between − 2.5 and − 3 V. A more precise measurement can be made, if necessary, by connecting an external d-c bias supply to the FET gate, adjusting the bias supply, and observing the exact value of pinch-off voltage on the oscilloscope.

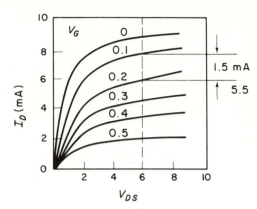

FIGURE 9–21 Measurement of FET transconductance (gain) using a curve tracer.

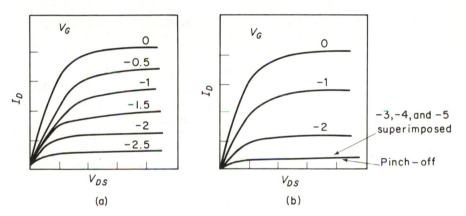

(a)   (b)

FIGURE 9–22 Determining FET pinch-off voltage using a curve tracer.

9–11–5. FET Zero-Temperature-Coefficient Point Tests

An interesting characteristic of all FETs is their ability to operate at a zero-temperature-coefficient (0TC) point. This means that if the gate–source is biased at a specific voltage and is held constant, the drain current does not vary with changes in temperature. This effect is shown in Fig. 9–23, which illustrates the I_D-V_{GS} curves of a typical FET for three different temperatures. Note that the three temperature curves intersect at a common point. If the FET is operated at this value of I_D and V_{GS} (shown as I_{DZ} and V_{GSZ}), 0TC operation results. That is, the drain current remains constant, even with extreme temperature changes.

185

$$I_{DZ} \approx I_{DSS}\left(\frac{0.63}{V_P}\right) \approx \left(\frac{0.4\,I_{DSS}}{V_P^2}\right)$$

$$V_{GSZ} \approx V_P - 0.63$$

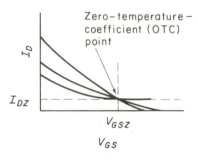

FIGURE 9-23 Zero temperature co-efficient for FETs.

As shown in Fig. 9–23, the values of I_D and V_{GS} that produce 0TC can be calculated by using datasheet curves or by equations. However, such values are typical approximations. A more practical method for determining I_{DZ} (the constant drain current at the 0TC point) requires a soldering tool, a coolant (a can of Freon), and a curve tracer. The curve tracer is adjusted to display a family of FET curves. By alternately bringing the soldering tool near the FET and spraying the FET with coolant, the voltage step of V_{GS} that *remains motionless* on the curve tracer can be observed. The I_D at this voltage step is the I_{DZ}. For example, if the 0.3-V step of Fig. 9–21 remains motionless, the I_{DZ} is 4 mA (and the 0.3-V V_{GS} produces the 0TC).

Typically, FETs with an I_{DSS} of about 10 to 20 mA have an I_{DZ} of less than 1 mA. Usually, the I_{DZ} increases as I_{DSS} increases (but not always, and not in proportion). For example, the I_{DZ} of a 50-mA FET often shows an I_{DZ} below 1 mA.

9–12. UJT INTERBASE CHARACTERISTIC TESTS

The interbase characteristics of UJTs are usually indicated by measurement of base 2 current I_{B2} as a function of interbase voltage V_{B2B1} and emitter current I_E. Such interbase characteristics are best measured on a sweep basis rather than with constant voltage and currents. This avoids the heating effects due to power dissipation.

A circuit for sweep test of UJT interbase characteristics is shown in Fig. 9–24. A constant-current pulse and a sweep voltage must be supplied by external test equipment. The constant-current pulse is applied to the emitter from

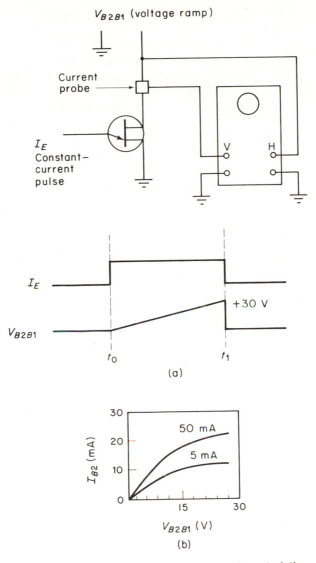

V_{B2B1} (voltage ramp)

Current probe

I_E
Constant–current pulse

I_E

V_{B2B1}

+30 V

t_0

t_1

(a)

I_{B2} (mA)

30

20

10

0

50 mA

5 mA

15

30

V_{B2B1} (V)

(b)

FIGURE 9-24 Measuring UJT interbase characteristics.

time zero t_0 to time t_1. Simultaneously, a voltage ramp going from 0 to 30 V is applied to base 2. Base 1 is grounded to complete the circuit. The current I_{B2} is measured with a current probe and applied to the vertical input of the oscilloscope. The voltage ramp, applied to base 2, is also applied to the oscilloscope horizontal input. Figure 9–24b shows typical interbase characteristics. As shown, the percentage increase in I_{B2} decreases with increasing emitter current and temperature.

9-13. UJT TRANSIENT CHARACTERISTIC TESTS

The transient characteristics of a UJT are usually not specified in the same way as for a two-junction transistor or FET. For example, switching times are usually not specified on a UJT datasheet. Instead, a parameter of f_{max} is given for most UJTs. The f_{max} indicates the maximum frequency of oscillation which can be obtained using the UJT in a specified relaxation oscillator circuit.

In some applications, such as critical timers, it may be of interest to determine turn-on and turn-off times associated with the UJT. The following paragraphs describe basic procedures for these measurements.

The circuit of Fig. 9–25a can be used to measure t_{on} and t_{off} for the case

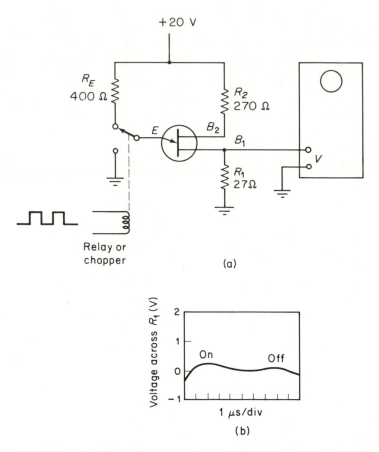

FIGURE 9-25 Measuring t_{on} and t_{off} of UJT with purely resistive load.

where the UJT emitter circuit is *purely resistive*. Typical switching-time values are $t_{on} = 1$ μs, $t_{off} = 2.5$ μs. The waveform observed at the base terminal when the UJT turns off is shown in Fig. 9–25 b.

The circuit of Fig. 9–26 can be used to measure t_{on} and t_{off} for the case where the UJT emitter circuit has both capacitance and resistance (as usually is the case). The test circuit of Fig. 9–26a is a *relaxation oscillator*, with turn-on and turn-off time being measured at the base 1 terminal. Typical turn-on and turn-off waveforms are shown in Fig. 9–26.

Turn-on time is measured from the start of turn-on to the 90% point. A typical turn-on time is 0.5 μs, with the capacitance of C_E shown in Fig. 9–26. An increase in C_E capacitance causes an increase in turn-on time. Turn-off time is measured from the start of turn-off to the 90% point, and is about 12 μs (due to the long discharge time of the capacitor). Turn-off time also increases with an increase in C_E capacitance. The effect of C_E capacitance on switching time is shown in Fig. 9–27.

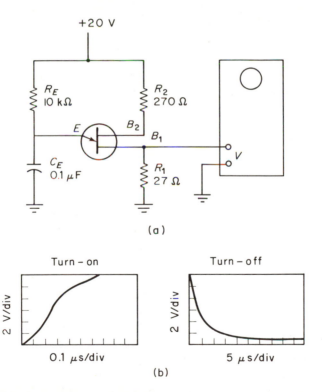

(a)

(b)

FIGURE 9-26 Measuring t_{on} and t_{off} of UJT with both capacitive and resistive loads.

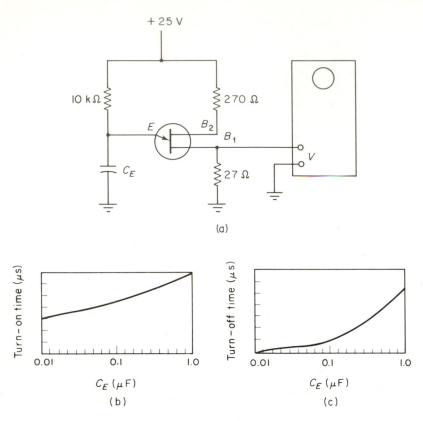

+ 25 V

10 kΩ   270 Ω

E B_2 B_1

C_E 27 Ω

V

(a)

(b)

Turn-on time (μs)

0.01   0.1   1.0

C_E (μF)

(c)

Turn-off time (μs)

0.01   0.1   1.0

C_E (μF)

FIGURE 9-27 Turn-on and turn-off times versus emitter capacitance in relaxation oscillator test circuit.

9–14. TESTING UJT EMITTER AND INTERBASE CHARACTERISTICS USING A CONVENTIONAL OSCILLOSCOPE

The circuit of Fig. 9–28 can be used to display either the emitter characteristic curves or the interbase characteristic curves on a conventional (non-curve tracer) oscilloscope. When switches are set to display the interbase curves, the meter indicates the emitter current. When the switches are set to display the emitter curves, the meter indicates the interbase voltage.

To avoid accidental burnout, it is important to set the variac and the d-c supply to zero before changing the switches or inserting the UJT in the circuit. If desired, external resistors may be inserted in series with the emitter, B_1, or B_2 to determine the characteristic of the UJT in a particular circuit.

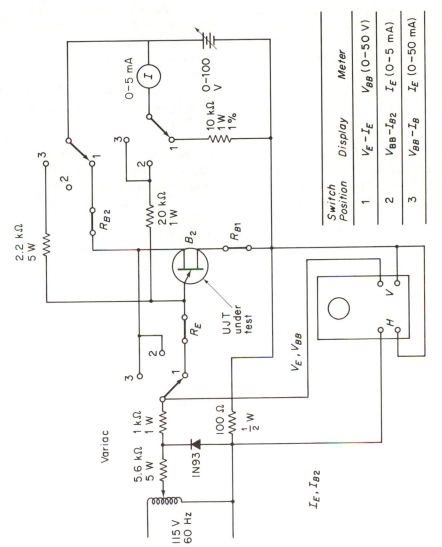

Switch Position	Display	Meter
1	$V_E - I_E$	V_{BB} (0–50 V)
2	$V_{BB} - I_{B2}$	I_E (0–5 mA)
3	$V_{BB} - I_B$	I_E (0–50 mA)

FIGURE 9-28 Emitter and interbase curve tracer test circuit using a conventional oscilloscope.

9-15. UJT INTRINSIC STANDOFF-RATIO TEST

The circuit of Fig. 9–29 can be used to measure the intrinsic standoff ratio, as well as the emitter-diode voltage V_D, of a UJT. In this circuit, the interbase voltage is swept by a 10-Hz oscillator, while the UJT oscillates at about 2 kHz in the basic relaxation oscillator configuration. The interbase voltage is applied to the horizontal axis of the oscilloscope. The emitter voltage is applied to one input of the oscilloscope vertical deflection amplifier. The R_2 potentiometer voltage is applied to the other input of the vertical deflection amplifier. With these connections, the oscilloscope pattern consists of a plot of V_{B2B1} on the horizontal axis against $V_P - KV_{B2B1} = V_D + (-K)V_{B2B1}$ on the vertical axis. K is equal to the fractional setting of potentiometer R_2. V_P is the upper envelope of the displayed emitter voltage.

During test, K (potentiometer R_2) is adjusted until the upper envelope of the display is horizontal. At this point, K or the fractional setting of R_2 is equal to the intrinsic standoff ratio. For example, if R_2 is 10 kΩ, and the resistance from the contact arm of R_2 to ground in 7 kΩ, the intrinsic standoff ratio is 0.7. If a precision potentiometer is used for R_2, the intrinsic standoff ratio can be measured with an accuracy of better than 0.05%.

With an R_2 adjusted to produce a horizontal envelope on the display, note the displacement of the upper envelope from the zero axis of the oscilloscope (in volts). This displacement is equal to V_D. If the oscilloscope

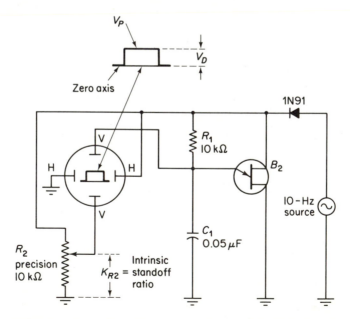

FIGURE 9-29   Measuring UJT intrinsic standoff ratio and V_D.

vertical screen can be read accurately, V_D can be measured within about 20 mV.

9-16. UJT TESTS USING A CURVE TRACER

Many curve tracers provide special circuits and connections for measurement of UJT characteristics. Most curve tracers can be converted by simple external circuits to test UJTs. For example, Fig. 9–30 shows the connections for conversion of a Tektronix 575 curve tracer to test UJTs. Note that for test of UJT interbase characteristics, the curve tracer terminals normally used for a two-junction transistor base is connected to the UJT emitter. For test of emitter curves, the UJT emitter is connected to the collector terminal of the test set. When displaying the emitter curves, the interbase voltage should not exceed 12 V because of the voltage limitation of the base-current step generator in the curve tracer.

The B&K Precision curve tracer can be used without modification to display a set of UJT curves, and to measure interbase resistance. The following paragraphs describe the basic steps.

Test	Emitter Curves	Interbase Curves
Circuit	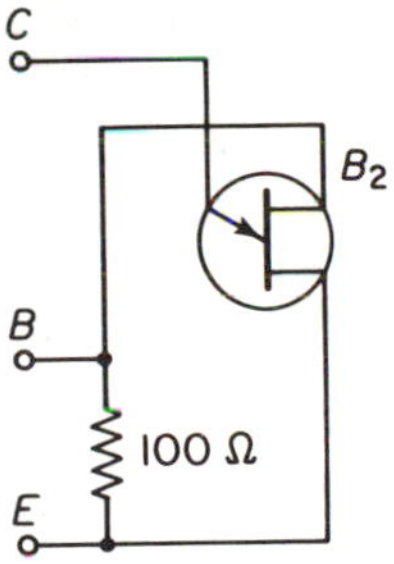	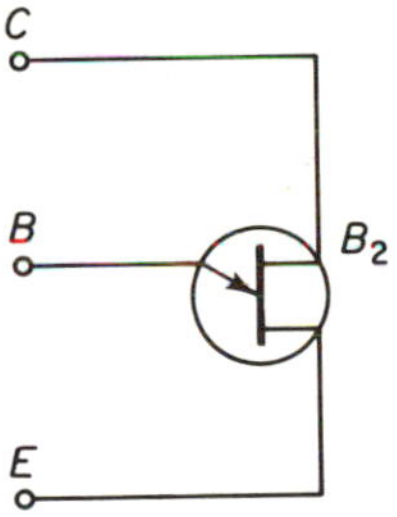
Collector sweep polarity	+	+
Base step polarity	+	+
Collector peak volts range	200 V	20 V
Collector limiting resistor	5 kΩ	500 Ω
Base–current step selector	20 mA/step	10 mA/step
Number of current steps	5	5
Vertical current range	2 mA/div I_E	2 mA/div I_{B2}
Horizontal voltage range	1 V/div V_E	2 V/div V_{BB}

FIGURE 9-30 Connections for conversion of Tektronix 575 curve tracer to test UJTs.

9-16-1. Displaying UJT Curves on a Curve Tracer

To display UJT curves on the curve tracer, connect the UJT to the tracer as shown in Fig. 9-31. As illustrated, base 1 is connected to the emitter jack, base 2 is connected to the base jack, and the (UJT) emitter is connected to the collector jack. (Note that base 1 and base 2 are interchangeable.)

Turn the curve tracer polarity switch to NPN. Increase the sweep voltage from zero until the trigger voltage is exceeded. This should produce the high emitter current spike on the oscilloscope. Set the step selector to the current per step position that produces the *most curves* on the display.

The curves may appear quite close together (compared to the display of a two-junction transistor or FET). This means that you must observe the display carefully to distinguish the curves. It may be helpful to spread out the display by increasing the horizontal sensitivity of the oscilloscope, or to use expanded sweep magnification (if available) at the area of interest.

With the test configuration as shown in Fig. 9-31, the step current of the curve tracer is applied from base 1 to base 2, and the sweep voltage is applied to the emitter. Thus, the sweep voltage is the UJT trigger voltage. Note that as the sweep voltage is slowly increased from the trigger threshold producing the first current spike, the other curves are added one by one. Thus, the emitter trigger voltage can be measured for each base-current step.

9-16-2. Measuring UJT Interbase Resistance with a Curve Tracer

Interbase resistance (shown as r_{BB} or R_{BB} on UJT datasheets) can be displayed using the connections shown in Fig. 9-32. As shown, base 1 and base 2 are connected to the curve tracer collector and emitter jacks, respectively. The UJT emitter is left open-circuited (no connection). The display

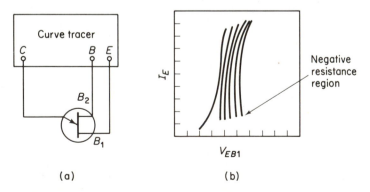

(a)   (b)

FIGURE 9-31   (a) Test connections and (b) typical display of UJT curves on a curve tracer.

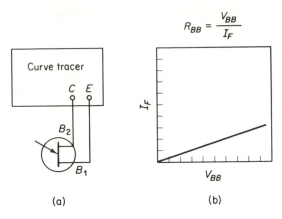

$$R_{BB} = \frac{V_{BB}}{I_F}$$

(a)   (b)

FIGURE 9-32 (a) Test connections and (b) typical display of UJT R_{BB} characteristic.

should be a linear trace as shown. The vertical scale of the display represents forward current (I_F), whereas the horizontal scale represents interbase voltage (V_{BB}). Interbase resistance equals interbase voltage divided by forward current, as shown in Fig. 9–32.

9-17. DIODE DYNAMIC TESTS

It is possible to test diodes using simple d-c test circuits (to measure leakage and voltage drop with a constant direct current). However, diodes do not usually operate this way in circuits. Instead, diodes are operated with alternating current, which tends to heat the diode junctions and change the characteristics. It is more realistic to test a diode under dynamic conditions. The following paragraphs describe two typical dynamic (alternating current) tests for diodes using an oscilloscope.

9-17-1. Dynamic Tests for Power Rectifier Diodes

Power rectifier diodes can be subjected to a dynamic test by using an oscilloscope to display and measure the current and voltage characteristics. To do so, a direct-current oscilloscope must be used, and both the horizontal and vertical channels must be *voltage-calibrated*. Usually, the horizontal channel of most oscilloscopes is time-calibrated. However, the horizontal channel can be voltage-calibrated by using the same procedure as for voltage calibration of the vertical channel (Chapter 6). In brief, the oscilloscope sweep circuit is disconnected from the input to the horizontal amplifier (the horizontal input is set to EXTERNAL), a reference voltage is applied to the horizontal amplifier input, and the horizontal gain or width control is set for some specific deflection or width (1 cm of horizontal width per volt, etc.). For best results, horizontal and vertical channels must be identical or nearly identical to eliminate any phase difference.

As shown in Fig. 9–33, a power rectifier diode is tested by applying a controlled a-c voltage across the anode and cathode through resistor R_1. The a-c voltage [set to the maximum rated peak inverse voltage (PIV) of the diode] alternately biases the anode positive and negative, causing both forward and reverse current to flow through R_1. The voltage drop across R_1 is applied to the vertical channel and causes the oscilloscope screen spot to move up and down. Vertical deflection is proportional to current through the diode under test. The vertical scale divisions can be converted directly to current when R_1 is made 1 Ω. For example, a 3-V vertical deflection indicates a 3-A current. If R_1 is 1000 Ω, the readout is in milliamperes.

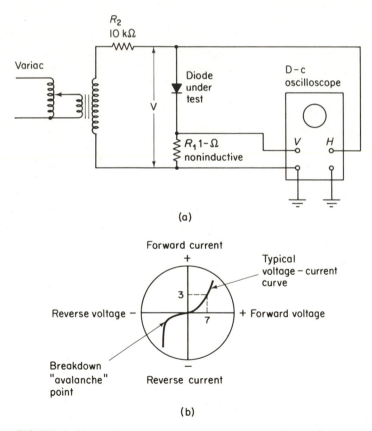

FIGURE 9–33   (a) Test connections and (b) typical display for power rectifier diode voltage–current characteristics. V = maximum rated voltage of diode under test.

The same voltage applied across the diode is applied to the horizontal channel (which has been voltage-calibrated), and causes the spot to move right or left. Horizontal deflection is proportional to voltage across the diode (neglecting the small voltage drop across R_1).

The combination of the horizontal (voltage) deflection and vertical (current) deflection causes the spot to trace out the complete current and voltage characteristics.

The basic test procedure is as follows:

1. Connect the equipment as shown in Fig. 9-33.
2. Place the oscilloscope in operation. Voltage-calibrate both the vertical *and horizontal* channels as necessary. The spot should be at the vertical and horizontal center with no signal applied to either channel.
3. Switch off the internal recurrent sweep of the oscilloscope. Set the sweep selector and sync selector (if any) of the oscilloscope to external. Leave the horizontal and vertical gain controls set at the (voltage) calibrate position as established in step 2.
4. Adjust the variac so that the voltage applied across the power diode under test is at (or just below) the maximum rated value (as determined from the diode datasheet).
5. Check the oscilloscope pattern against the typical curves of Fig. 9-33 and/or against the diode specifications. The curve of Fig. 9-33 is a typical response pattern. That is, the forward current (deflection above the horizontal centerline) increases as forward voltage (deflection to the right of the vertical centerline) increases. Reverse current increases only slightly as reverse voltage is applied, unless the breakdown or "avalanche" point is reached. In conventional (nonzener) diodes, it is desirable (if not mandatory) to operate considerably below the breakdown point. Some diodes will break down if operated in the reverse condition for any length of time.
6. Compare the current and voltage values against the values specified in the diode datasheet. For example, assume that a current of 3 A flows with 7 V applied. This can be checked by measuring along the horizontal scale to the 7-V point, then measuring from that point up (or down) to the trace. The 7-V (horizontal) point should intersect the trace at the 3-A (vertical) point as shown.

9-17-2. Dynamic Tests for Small-Signal Diodes

The procedures for checking the current–voltage characteristics of a signal diode are the same as for power-rectifier diodes. However, there is one major difference. In a small-signal diode, the ratio of forward voltage to

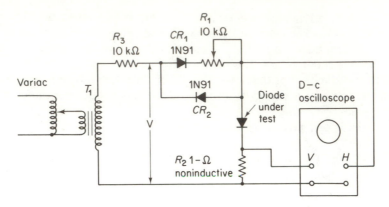

FIGURE 9-34 Test connections for measurement of signal diode voltage–current characteristics. V = maximum rated voltage of diode under test.

reverse voltage is usually quite large. A test of forward voltage at the same amplitude as the rated reverse voltage will probably damage the diode. On the other hand, if the test voltage is lowered for both forward and reverse directions, the voltage is not a realistic value in the reverse direction.

Under ideal conditions, a signal diode should be tested with a low-value forward voltage and a high-value reverse voltage. This can be done using a circuit shown in Fig. 9–34. The circuit of Fig. 9–34 is essentially the same as that of Fig. 9–33 (for power diodes), except that diodes CR_1 and CR_2 (Fig. 9–34) are included to conduct on alternate half-cycles of the voltage across transformer T_1. Rectifiers CR_1 and CR_2 are chosen for a linear amount of conduction near zero.

The variac is adjusted for maximum rated reverse voltage across the diode under test, as applied through CR_2, when the upper secondary terminal of T_1 goes negative. This applies the full reverse voltage.

Resistor R_1 is adjusted for maximum rated forward voltage across the diode, as applied through CR_1, when the upper secondary terminal of T_1 goes positive. This applies a forward voltage limited by R_1.

With resistor R_1 properly adjusted, perform the current–voltage check as described for power diodes (Sec. 9–17–1).

9-18. DIODE SWITCHING TESTS

Diodes to be used in pulse or digital work must be tested for switching characteristics. The single most important switching characteristic is *recovery time*. When a reverse-bias pulse is applied to a diode, there is a measureable time delay before the reverse current reaches its steady-state value. This delay period is listed as the recovery time (or some similar term) on the diode datasheet.

The duration of recovery time sets the minimum width for pulse with which the diode can be used. For example, if a 5-μs reverse voltage pulse is applied to a diode with a 10-μs recovery time, the pulse will be distorted.

An oscilloscope having a wide frequency response and good transient characteristics can be used to check the high-speed switch and recovery time of diodes. The oscilloscope vertical channel must be voltage-calibrated in the normal manner, and the horizontal channel must be time-calibrated (rather than sweep-frequency-calibrated). As discussed in Chapter 7, most laboratory oscilloscopes are time-calibrated, and some shop oscilloscopes are frequency-calibrated.

As shown in Fig. 9–35, the diode is tested by applying a forward-bias current from a direct-current supply, adjusted by R_1 and measured by M_1. The negative portion of the square wave out from the square-wave generator is developed across R_3. The square wave switches the diode voltage rapidly to a high negative value (reverse voltage). However, the diode does not cut off immediately. Instead, a steep transient voltage is developed by the high momentary current flow. The reverse current falls to its steady-state value when the carriers are removed from the junction. This produces the approximate waveform shown in Fig. 9–35.

Both forward and reverse currents are passed through resistor R_3. The voltage drop across R_3 is applied through emitter follower Q_1 to the oscilloscope vertical channel. The coaxial cable provides some delay so that the complete waveform is displayed. CR_1 functions as a clamping diode to keep the R_4 voltage at a level safe for the oscilloscope.

The time interval between the negative peak and the point at which the reverse current has reached the low, steady-state value is the diode recovery time. Typically, this time is in the order of a few nanoseconds for a signal diode.

The test procedure is as follows:

1. Connect the equipment as shown in Fig. 9–35.
2. Place the oscilloscope in operation as described in the instruction manual.
3. Switch on the oscilloscope internal recurrent sweep. Set the sweep selector and sync selector to internal.
4. Set the square-wave generator to a repetition rate of 100 kHz or specified in the diode datasheet.
5. Set R_1 for the specified forward test current as measured on M_1.
6. Increase the square-wave generator output level (amplitude) until a pattern appears.

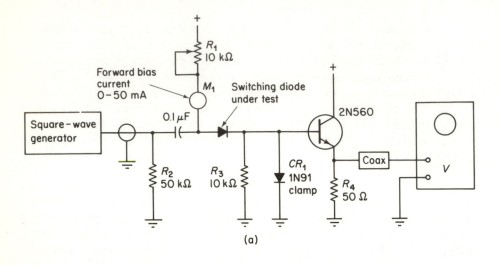

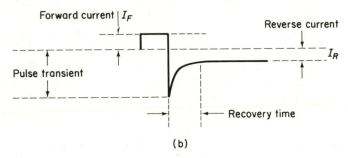

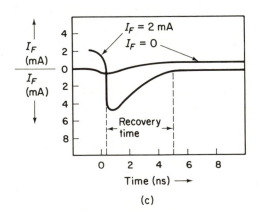

FIGURE 9-35 (a) Test circuit, (b) theoretical display, and (c) practical display for switching (recovery) time of diodes.

7. If necessary, readjust the sweep and sync controls until a single sweep is shown.
8. Measure the recovery time along the horizontal (time-calibrated, probably in nanoseconds) axis.

9–19. ZENER DIODE TESTS

The test of a zener diode is similar to that of a power rectifier or signal diode. The forward voltage drop test for a zener is identical to that of a conventional diode. A reverse leakage test is usually not required, since a zener goes into the avalanche condition when sufficient reverse voltage is applied. In place of a reverse leakage test, a zener diode is tested to determine the point at which avalanche occurs (establishing the zener voltage across the diode).

The procedures and circuit for dynamic test of zener diodes are similar to those for dynamic test of conventional diodes. As shown in Fig. 9–36, the zener diode is tested by applying a controlled a-c voltage across the anode and cathode through resistors R_1 and R_2. The a-c voltage (set to some value above the zener voltage) alternately biases the anode positive and negative, causing both forward the reverse current to flow through R_1 and R_2.

The voltage drop across R_2 is applied to the vertical channel and causes the screen spot to move up and down. Vertical deflection is proportional to current. The vertical scale divisions can be converted directly to current when R_2 is made 1 Ω. For example, a 3-V vertical deflection indicates a 3-A current.

The same voltage applied across the zener (taken from the junction of R_1 and the zener under test) is applied to the horizontal channel (which has been voltage-calibrated) and causes the spot to move right or left. Horizontal deflection is proportional to voltage. The combination of the horizontal (voltage) deflection and vertical (current) deflection causes the spot to trace out the complete current and voltage characteristics.

The test procedure is as follows:

1. Connect the equipment as shown in Fig. 9–36.
2. Place the oscilloscope in operation as described in the instruction manual. Voltage-calibrate both the vertical and horizontal channels as necessary. The spot should be at the vertical and horizontal center with no signal applied to either channel.
3. Switch off the internal recurrent sweep. Set sweep selector and sync selector controls to external. Leave the horizontal and vertical gain controls set at the (voltage) calibrate position established in step 2.
4. Adjust the variac so that the voltage applied across the zener diode, and resistors R_1 and R_2 in series, is greater than the rated zener voltage.

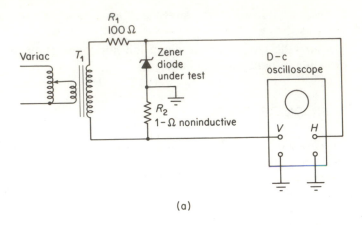

(a)

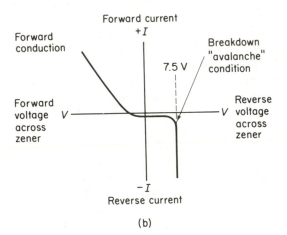

(b)

FIGURE 9-36 (a) Test connections and (b) typical display for zener diodes.

5. Check the oscilloscope pattern against the typical curves of Fig. 9-36 and/or against the diode specifications. The curve of Fig. 9-36 is a typical response pattern. That is, the forward current increases as forward voltage increases. Reverse (or leakage) current increases only slightly as reverse voltage is applied until the avalanche or zener voltage is reached. Then the current increases rapidly (thus the term "avalanche").

6. Compare the current and voltage values against the values specified in the zener diode datasheet. For example, assume that avalanche current should occur when the reverse voltage (zener voltage) reaches 7.5 V. This can be checked by measuring along the horizontal scale to the 7.5-V point.

9-20. TUNNEL DIODE TESTS

The single most important test of a tunnel diode is the negative-resistance characteristic. The most effective test of a tunnel diode is to display the entire forward voltage and current characteristics on an oscilloscope, permitting the valley and peak voltages, as well as the valley and peak currents, to be measured simultaneously.

    A d-c oscilloscope is required for test of a tunnel diode. Both the vertical and horizontal channels must be voltage-calibrated. Also, the horizontal and vertical channels must be identical, or nearly identical, to eliminate any phase difference. As shown in Fig. 9–37, the tunnel diode is tested by applying a

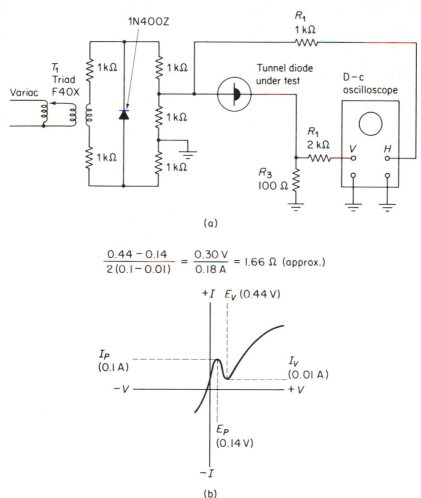

$$\frac{0.44 - 0.14}{2\,(0.1 - 0.01)} = \frac{0.30\,\text{V}}{0.18\,\text{A}} = 1.66\,\Omega \ \ (\text{approx.})$$

(b)

FIGURE 9-37 (a) Negative resistance test connections and (b) typical display for tunnel diodes.

controlled d-c voltage across the diode through resistor R_3. This d-c voltage is developed by rectifier CR_1 and is controlled by the variac. Current through the tunnel diode also flows through R_3. The voltage drop across R_3 is applied to the vertical channel and causes the spot to move up and down. Thus, vertical deflection is proportional to current. Vertical scale divisions can be converted directly to a realistic value of current when R_3 is made 100 Ω. For example, a 3-V vertical deflection indicates 30 mA.

The same voltage applied across the tunnel diode is applied to the horizontal channel (which has been voltage-calibrated) and causes the spot to move from left to right. For a tunnel diode, the horizontal and vertical zero-reference point (no-signal spot position) is at the lower left of the screen rather than in the center. The horizontal deflection is proportional to voltage. The combination of the horizontal (voltage) deflection and vertical (current) deflection causes the spot to trace out the complete negative-resistance characteristic.

The test procedure is as follows:

1. Connect the equipment as shown in Fig. 9–37.
2. Place the oscilloscope in operation as described in the instruction manual. Voltage-calibrate both the vertical and horizontal channels as necessary. The spot should be at the lower left-hand side of center with no signal applied to either channel.
3. Switch off the internal recurrent sweep. Set the sweep selector and sync selector controls to external. Leave the horizontal and vertical gain control set at the (voltage) calibration position, as established in step 2.
4. Adjust the variac so that the voltage applied across the tunnel diode under test is the maximum rated forward voltage (or slightly below). This can be read across the voltage-calibrated horizontal axis.
5. Check the oscilloscope pattern against the curve of Fig. 9–37 or against the tunnel diode datasheet.
6. The following equation can be used to obtain a *rough* approximation of negative resistance in tunnel diodes.

$$\text{negative resistance} = \frac{E_V - E_P}{2(I_P - I_V)}$$

where E_V = valley voltage
         I_V = valley current
         E_P = peak voltage
         I_P = peak current

Figure 9–37 shows some typical tunnel diode voltage–current values, and the calculations for the approximate negative resistance.

9-21. SIGNAL AND POWER DIODE TEST USING A CURVE TRACER

Signal and power (rectifier) diodes conduct easily in one direction and are nonconducting in the opposite direction. These properties may be tested and observed with a curve tracer and oscilloscope. For testing diodes, the pulsating d-c sweep voltage is applied across the diode, and the diode current–voltage characteristics are plotted on the oscilloscope screen. The step current or step voltage signals used for testing transistors and FETs are not used in diode testing, and the curve tracer step selector has no effect on test results.

9-21-1. Basic Diode Test Connections

The diode under test can be plugged into the collector and emitter pins of the transistor socket, or directly into the collector and emitter jacks, depending on the type of curve tracer. Test leads can also be run from the connector and emitter jacks to the terminals of the diode under test.

Since the polarity of the sweep voltage can be reversed with the curve tracer polarity switch, the diode may be inserted into the socket without observing polarity. Of course, diodes connected with one polarity produce an oscilloscope display that deflects to the right and upward from the starting point, whereas the opposite polarity produces a display that deflects downward and to the left from the starting point. For a consistent display, connect the cathode of the diode to the emitter jack as shown in Fig. 9–38. With this polarity connection, and the polarity switch set to NPN, the diode is forward-biased, and should produce a display similar to that of Fig. 9–38. With the

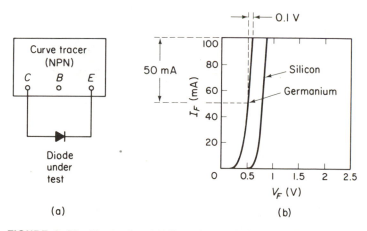

(a)

(b)

FIGURE 9-38 Measuring (a) diode forward bias using (b) a curve tracer.

same polarity connection, and the polarity switch set to PNP, the diode is reverse-biased, and should produce a display similar to that of Fig. 9–39.

9-21-2. Diode Forward-Bias Display on a Curve Tracer

To display forward bias, connect the diode and operate the curve tracer controls as shown in Fig. 9–38. For realistic values, the oscilloscope horizontal sensitivity should be calibrated at some low voltage. For a typical diode, a horizontal sensitivity of 0.5 V/division (or preferably less) is necessary to obtain any degree of accuracy in the voltage reading. Most oscilloscopes used with curve tracers can be calibrated for a sensitivity of 0.25 V/division in the horizontal axis.

When testing diodes as described here, only one curve is displayed, not a family of curves as displayed for transistors and FETs. Both *forward voltage drop* and *dynamic resistance* can be measured using the connections of Fig. 9–38. As shown, no current flows until the applied voltage exceeds the forward voltage drop. Typically, the forward voltage drop of germanium diodes is in the range 0.2 to 0.4 V, whereas the forward voltage drop of silicon diodes is about 0.5 to 0.7 V. Above the forward voltage point, current increases rapidly with an increase in forward voltage. As shown, the current increases more rapidly and the "elbow" has a sharper bend for silicon diodes than for germanium diodes.

The dynamic resistance of a diode equals the change in forward voltage divided by the change in forward current. For example, the germanium diode curve of Fig. 9–38 shows an increase of about 50 mA in forward current for an increase of about 0.1 V of forward voltage. This indicates a dynamic resistance of about 2 Ω. Germanium diodes, with more slope to their curves as shown, have higher dynamic resistance and silicon diodes.

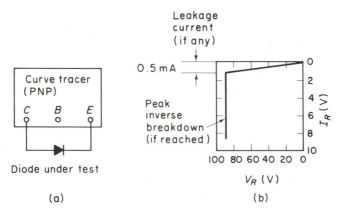

FIGURE 9-39 Measuring (a) diode reverse bias using (b) a curve tracer.

There is no need to increase the curve tracer sweep voltage control setting beyond that which gives a full-scale vertical presentation, although there is generally very little danger that a higher setting will do harm. Typically, the oscilloscope vertical sensitivity control may be set to the 1/mA/div position for examining the low-current characteristics, or to a lower sensitivity position (such as 10 mA/div) for observing a wider range of forward current conduction.

9–21–3.   Diode Reverse-Bias Display on a Curve Tracer

To display reverse bias, connect the diode and operate the curve tracer controls as shown in Fig. 9–39. When the polarity switch is first set to the opposite position as used for the forward-bias tests, there should be only a horizontal line displayed. The oscilloscope centering must be readjusted because the polarity reversal causes the trace to move off-screen.

Leakage current is easier to check with a much higher voltage than is used for forward-bias testing. Typically, the oscilloscope can be calibrated for 10 V/division, which allows display of the test up to 100 V. Keep in mind that the diode must never be operated at reverse voltages greater than the peak inverse voltage (or breakdown voltage) for any prolonged period of time. To do so can damage the diode. Note that leakage current is displayed as a sharp vertical drop. Figure 9–39 shows a leakage current of about 0.5 mA at a breakdown voltage of 90 V, with corresponding less leakage at lower reverse voltages.

9–22.   ZENER DIODE TEST USING
A CURVE TRACER

When a curve tracer is used, the procedure for testing zener diodes is almost the same as for testing signal and rectifier diodes. In fact, the forward characteristics of the diodes are essentially identical and the test procedures would be the same, except that forward voltage measurements are seldom used for zener diodes, since zeners are designed to be used in the reverse voltage breakdown mode. In this mode, a large change in reverse current occurs while the zener voltage remains nearly constant. Because of this characteristic, zener diodes are most often used as voltage regulators.

The zener voltage value (reverse voltage breakdown value) may be measured with the curve tracer and oscilloscope set up as described for reverse voltage measurement of signal and power diodes (Sec. 9–21 and Fig. 9–39). To get the most accurate voltage reading possible, calibrate the full-scale oscilloscope horizontal sensitivity to a convenient value slightly above the zener voltage. For example, for a 6-V diode, calibrate full scale at 10 V.

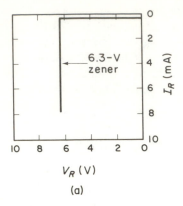
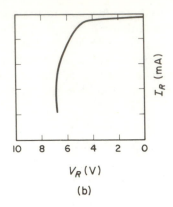

FIGURE 9-40 Typical zener diode curve tracer displays: (a) sharp zener knee; (b) sloppy zener knee.

Be sure the curve tracer polarity switch is set to display the reverse voltage condition. Increase the curve tracer sweep voltage control setting to display the zener region as shown in Fig. 9-40. No reverse current should flow until the reverse breakdown voltage value is reached. At that point, there should be a very sharp "elbow" or "knee" and a very vertical current trace. If the knee is rounded or the vertical current trace has a measurable voltage slope, the zener diode is probably defective. Read the zener voltage from the display.

9-23. TUNNEL DIODE TEST USING A CURVE TRACER

The characteristics of tunnel diodes may be measured with a curve trace by connecting the diodes between the collector and emitter jacks as shown in Fig. 9-41. If the cathode cannot be readily identified, try both positions of the POLARITY switch. The display shown in Fig. 9-41a is obtained when the tunnel diode cathode is connected to the emitter jack, and the curve tracer polarity switch is set to NPN.

Calibrate the oscilloscope horizontal amplifier for high sensitivity, such as 0.1 V/division. It may be necessary to use sweep magnification or horizontal scale expansion (if the oscilloscope is capable of such operation) to get this sensitivity. Set the curve tracer sweep voltage control to a low value so that the voltage just sweeps through the tunnel region (typically less than 0.5 V). Set the oscilloscope vertical sensitivity as required for the largest possible on-scale display.

Several characteristics of the tunnel diode can be measured directly from the display, as discussed in Sec. 9-20.

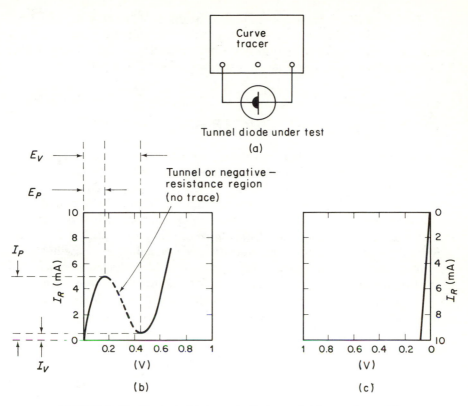

FIGURE 9-41 Typical (b) tunnel diode and (c) tunnel rectifier (reverse characteristics) curve tracer displays.

9-23-1. Tunnel Rectifier Test Using a Curve Tracer

Tunnel rectifiers are similar to tunnel diodes, but do not use the negative resistance characteristic in operation. The region that tends to tunnel is more resistive in tunnel rectifiers, and peak current is not as pronounced. Because of this characteristic, a high front-to-back ratio at low voltages allows the tunnel rectifier to be used as a very low signal voltage rectifier. The tunnel rectifier conducts very easily in one direction with very little voltage drop (actually the opposite direction from conventional diodes insofar as the N and P material is concerned). Thus tunnel rectifiers are sometimes called *back diodes*.

The reverse direction in a tunnel rectifier (direction that tends to tunnel) is resistive at low voltage values, but conducts readily at voltages that approximate the forward drop of a conventional diode. Thus, the peak voltage of the signal to be rectified should not exceed that of the resistive region.

Tunnel rectifier characteristics should be displayed as described for tunnel diodes. Tests using both polarities of sweep voltage are required to examine the forward versus reverse condition characteristics. The display of Fig. 9–41b shows typical reverse characteristics of a tunnel rectifier.

9-24. *BASIC CONTROL RECTIFIER AND THYRISTOR TESTS*

As in the case of diodes and transistors, control rectifiers are subjected to many tests during manufacture. Few of these test need be duplicated in the field. One of the simplest and most comprehensive tests for a control rectifier (sometimes called a thyristor) is to operate the device in a circuit that simulates actual circuit conditions (typically, alternating current and an appropriate load at the anode, alternating current or a pulse signal at the gate), and then measure the resulting conduction angle on a dual-trace oscilloscope.

With such a test procedure, the trigger and anode voltages, as well as the load current, can be adjusted to normal (or abnormal) dynamic operating conditions, and the results noted. For example, the trigger voltage can be adjusted over the supposed minimum and maximum trigger levels. Or the trigger can be removed, and the anode voltage raised to the actual breakover. The condition angle method should test the most important characteristics of a control rectifier, with the possible exception of turn-on and turn-off (which are also discussed in this section).

9-24-1. *Conduction Angle Test*

A dual-trace oscilloscope can be used to measure the conduction angle of a control rectifier or thyristor. As shown in Fig. 9–42, one trace of the oscilloscope displays the anode current, while the other trace displays the trigger voltage. Both traces must be voltage-calibrated. The anode load current is measured through a 1-Ω noninductive resistor. The voltage developed across this resistor is equal to the current. For example, if a 3-V indication is obtained on the oscilloscope screen, a 3-A current is flowing in the anode circuit. The trigger voltage is read out directly on the other oscilloscope trace. Note that a diode is placed in the trigger circuit to provide a pulsating d-c trigger. This can be removed if desired. Since the trigger is synchronized with anode current (both are obtained from the same source), the portion of the trigger cycle in which anode current flows is the conduction angle.

The test procedure is as follows:

1. Connect the equipment as shown in Fig. 9–42.
2. Place the oscilloscope in operation as described in the applicable in-

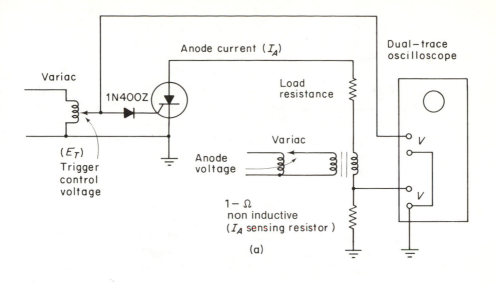

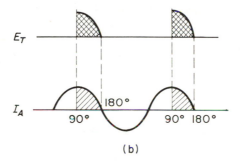

FIGURE 9-42 Basic test circuit for measurement of SCR or thyristor conduction angle.

struction manual. Switch on the internal recurrent sweep. Set the sweep selector and sync selector to internal.

3. Apply power to the control rectifier. Adjust the trigger voltage, anode voltage, and anode current to the desired levels. Anode voltage can be measured by temporarily moving the oscilloscope probe (normally connected to measure gate voltage) to the anode.

4. Adjust the oscilloscope sweep frequency and sync controls to produce two or three stationary cycles of each wave on the screen.

5. On the basis of one conduction pulse equaling 180°, determine the angle of anode current flow, by reference to the trigger voltage trace. For example, in the display of Fig. 9–42, anode current starts to flow at 90° and stops at 180°, giving a conduction angle of 90°.

NOTE

If the device under test is a triac (or similar unit such as an SBS), there is a conduction display on both half-cycles.

6.   To find the minimum or maximum required trigger level, vary the trigger voltage from zero across the supposed operating range, and note the level of trigger voltage when anode conduction starts.

7.   To find the breakdown voltage, remove the trigger voltage, and move the oscilloscope probe to the anode. Increase the anode voltage until conduction starts, and note the anode voltage level.

9-24-2.   Turn-On and Turn-Off (Recovery) Time Tests

Figure 9–43 shows a circuit capable of measuring both turn-on and turn-off (recovery) time. The circuit inductances must be kept to a minimum by using short connections, thick wires, and closely spaced return loops or wiring on a grounded chassis. External pulse sources must be provided for the circuit. These pulses are applied to transformers T_1 and T_2, and serve to turn the device under test on and off. The pulses can come from any source, but

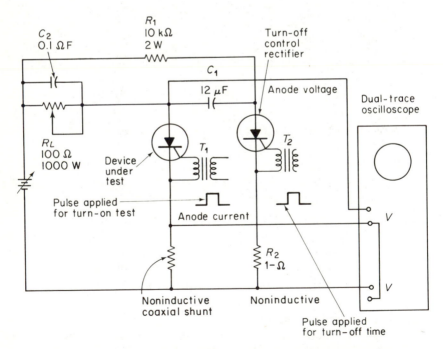

FIGURE 9-43   Test circuit for measurement of control rectifier and thyristor turn-on and turn-off (recovery) times.

should be of the amplitude, duration, and repetition rate that correspond to the normal operating conditions of the device under test.

When a suitable gate pulse is applied to transformer T_1, the device under test is turned on. Load current can be set by resistor R_L. A predetermined time later, the turn-off control rectifier is turned on by a pulse applied to T_2. This places capacitor C_2 across the device under test, applying a reverse bias, and turning the device under test off.

Any oscilloscope capable of a 10-μs sweep can be used for viewing both the turn-on and turn-off action. The oscilloscope is connected with the vertical input across the device under test. Turn-on time is displayed when the oscilloscope is triggered with the gate pulse applied to the device under test. Turn-off time is displayed when the oscilloscope is triggered with the gate pulse applied to the turn-off control rectifier.

The actual spacing between the turn-on and turn-off pulses is usually not critical. However, a greater spacing causes increased conduction, and heats the junction. Since operation of control rectifiers is temperature-dependent, the rise in junction temperature must be taken into account for accurate test results. (Both turn-on and turn-off times increase with an increase in junction temperature.)

Figure 9–44 shows turn-on action. Turn-on time is equal to delay time (t_d), plus rise time (t_r). Following the beginning of the gate pulse, there is a short delay before appreciable load current flows. Delay time is the time from the leading edge of the gate-current pulse (beginning of oscilloscope sweep) to the point of 10% load-current flow. (Delay time can be decreased by overdriving the gate.)

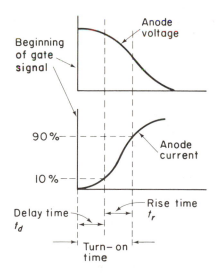

FIGURE 9–44   Definitions of turn-on time (delay time plus rise time).

Rise time (t_r) is the time the load current increases from 10% to 90% of the total value. Rise time depends upon load inductance, load-current amplitude, junction temperature, and to a lesser extent, upon anode voltage. The higher the inductance and load current, the longer the rise time. An increase in anode voltage tends to decrease the rise time. Capacitor C_2 (Fig. 9–43) tends to counter the load inductance, thus lessening the rise time.

By triggering the oscilloscope with the gate pulse applied to the device under test, the sweep starts at the gate pulse leading edge. Thus, the oscilloscope presentation shows the anode voltage from this point on. In noninductive circuits, when the anode voltage decreases to 90% of initial value, this time is equal to 10% of the load current, and is thus equal to the delay time.

With the oscilloscope set at 10-μs sweep, each square represents 1 μs. The delay time is read directly by counting the number of divisions on the oscilloscope screen. If the circuit is noninductive, the decrease from 90% to 10% of the anode voltage is approximately equal to the load current increase from 10% to 90%. The time this takes is equal to the rise time. The total time from zero time (start of oscilloscope sweep) to this 10% of the anode voltage is equal to the turn-on time. Therefore, turn-on time is determined by counting the number of divisions from the start of the oscilloscope sweep to the 90% anode load current.

Figure 9–45 shows the reverse current and reverse recovery (turn-off) action of the device under test. Turn-off time is the time necessary for the device under test to turn off and recover its forward blocking ability. The reverse recovery time (t_h) is the length of the interval between the time the forward

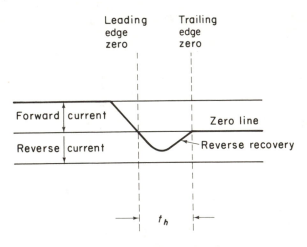

Reverse recovery time

FIGURE 9–45   Definition of reverse recovery (turn-off) time.

current falls to zero when going reverse, and the time the current returns back to zero from the reverse direction.

In Fig. 9–43, the time available for turn-off action is determined by the value of capacitor C_1 and resistor R_2. Decreasing the value of C_1 decreases the time the device under test is reverse-biased. The resistor R_2 limits the magnitude of the reverse current. The shape of the reverse voltage and current pulses are determined by the capacitor–resistor discharge. At the end of the reverse pulse, forward voltage is reapplied. Having turned off, the device under test blocks forward voltage, and no current can flow.

Figure 9–45 shows the reverse current pulse. With the oscilloscope set for a 20-μs sweep, the value of reverse recovery time (turn-off time) can be measured by counting off the divisions from the zero point on the leading edge of the reverse current pulse, to the zero point on the trailing edge, as shown.

9–25.   CONTROL RECTIFIER AND THYRISTOR TESTS USING A CURVE TRACER

A curve tracer can be used to test all important characteristics of control rectifiers and thyristors. This section describes a series of tests for SCRs, triacs, and diacs.

9–25–1.   SCR Tests Using a Curve Tracer

The following characteristics of an SCR can be displayed and measured with a curve tracer: forward blocking voltage, reverse blocking voltage, leakage current, holding current, forward voltage drop for various forward currents, and gate trigger voltage for various forward voltages.

9–25–2.   SCR Test Connections

For all measurements except gate trigger voltage, the SCR should be connected to the curve tracer as follows (Fig. 9–46): SCR cathode to emitter jack, or emitter pin of socket; SCR anode to collector jack, or collector pin of socket; SCR gate to base jack, or base pin of socket.

9–25–3.   Forward Blocking Voltage

To measure forward blocking voltage, set the curve tracer STEP SELECTOR to the I_{CES} position. This shorts the gate and cathode to satisfy the zero-gate-current requirement. Set the curve tracer POLARITY switch to NPN. Increase the curve tracer SWEEP VOLTAGE until the SCR fires or triggers. When this occurs, the anode current suddenly increases, and the

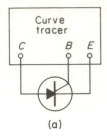

(a)

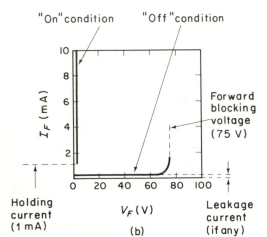

(b)

Holding current (1 mA)

V_F (V)

Leakage current (if any)

FIGURE 9-46 Measuring SCR forward blocking voltage and holding current using a curve tracer.

anode voltage drops to near zero, as shown in Fig. 9–46. In the "off" condition, the trace is horizontal until the forward blocking voltage is reached. In the "on" condition, the trace is vertical.

Read the highest anode voltage point in the display (75 V in Fig. 9–46). This is the maximum forward blocking voltage. Any anode current at anode voltage below the "firing" point is forward leakage current, and can be read directly from the display. (If the horizontal "off" condition trace is above zero, it is a result of forward leakage current.)

9-25-4. Reverse Blocking Voltage

The procedure for measuring reverse blocking voltage is the same as for measuring forward blocking voltage, except that the POLARITY switch is set to PNP. The voltage at which voltage breakdown occurs, which is a sudden increase in anode current, is the reverse-blocking-voltage value. Any anode current at voltages below breakdown is reverse leakage current, and can be read directly from the display.

216

9-25-5.   Holding Current

Using the same procedure as described for the forward blocking voltage test (Fig. 9-46), note the lowest current displayed for the "on" condition (1 mA in Fig. 9-46). This is the holding current. The measurement can also be made with the step selector in one of the current per step positions so that less sweep voltage is required to place the SCR in the "on" condition.

9-25-6.   Forward Voltage Drop

The forward voltage drop during the "on" condition at various forward current levels may be measured by increasing the horizontal sensitivity of the oscilloscope, and displaying a low-voltage portion of the forward voltage. Increase the step selector current per step setting so that sweep voltage may be reduced. The vertical sensitivity may also be reduced to a typical 10 mA/division for a greater range of voltage versus current. Read the forward voltage drop (for a given forward current) directly from the display.

9-25-7.   Gate Trigger Voltage

The turn-on point of an SCR depends on the forward voltage and gate voltage. As gate voltage is increased, less forward voltage is required to switch on the SCR. Conversely, as forward voltage is increased, less gate voltage is required to switch on the SCR. Gate trigger voltage values can be measured by connecting a d-c bias supply to the gate terminal of the SCR, as shown in Fig. 9-47. The bias supply reference must be connected to the emitter jack or to the cathode of the SCR. Otherwise, the curve tracer is set up as for forward blocking voltage measurement. Two types of measurements can be made.

1. Set the sweep voltage to a specified forward anode-cathode voltage, and increase the d-c bias supply until the SCR switches on. Measure the value of gate voltage (on meter M_1) at which switching occurs.
2. Set the d-c bias supply to a specified gate voltage, and increase the sweep voltage until the SCR switches on. Read the peak value of sweep voltage (on the oscilloscope display) that is required to produce triggering.

9-25-8.   Triac Tests Using a Curve Tracer

Triacs may be tested on a curve tracer exactly as SCRs, except that forward tests should be repeated for both directions. Also, there is no reverse-blocking-voltage measurement, since a triac normally conducts in both directions.

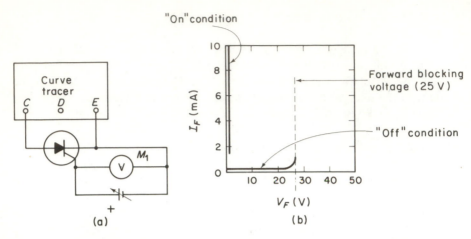

FIGURE 9-47 Measuring SCR turn-on point using a curve tracer and variable d-c supply.

9-25-9. Diac Tests Using a Curve Tracer

Breakdown voltage, leakage current, and holding current of a diac can be tested using a curve tracer. The procedures are the same as for an SCR. The diac is connected between the collector and emitter terminals of the curve tracer, and the sweep voltage is increased until breakdown occurs. The breakdown voltage holding current, and leakage, can be read from the oscilloscope display. The breakdown voltage is the highest voltage in the "off" condition. Leakage current is any deflection of the horizontal "off" condition trace from zero. Holding current is the lowest current in the "on" condition.

9-26. USING AN OSCILLOSCOPE AS THE NULL DETECTOR OF AN A-C BRIDGE

It is possible to use an oscilloscope, instead of a voltmeter, as the null detector of an a-c bridge. An oscilloscope offers the advantage of indicating both reactive balance and resistive balance.

As shown in Fig. 9–48, the source voltage is applied to the bridge input and to the oscilloscope horizontal channel through a phase-shift network. The bridge output is applied to the oscilloscope vertical channel.

1. Connect the equipment as shown in Fig. 9–48.
2. Place the oscilloscope in operation (Chapter 5). Switch off the internal recurrent sweep. Set sweep selector and sync selector to external.

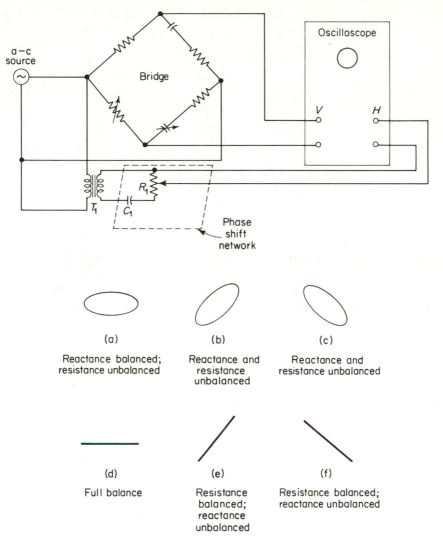

FIGURE 9-48 Using an oscilloscope as the null detector of an a-c bridge.

3. Connect the unknown test component (inductor, capacitor, resistor, etc.) to the appropriate bridge terminals.

4. Leave the bridge unbalanced. Adjust R_1 until the oscilloscope pattern becomes an ellipse. If necessary, adjust the horizontal and vertical gain controls to produce an ellipse of suitable size on the oscilloscope screen.

5. Note the position of the ellipse. When the reactance is balanced

(reactance null), the ellipse will be horizontal, as shown in Fig. 9–48a. If the ellipse is slanted to right or left, as shown in Fig. 9–48b or c, the reactance is unbalanced. Adjust the bridge reactance control until the ellipse is horizontal (Fig. 9–48a). Read the reactance of the unknown component from the bridge reactance control.

6. Adjust the bridge resistance control until the ellipse is closed into a straight line.

7. When both the reactance and resistance are balanced, a straight horizontal line is obtained, as shown in Fig. 9–48d. If only the resistance is balanced, the trace will be a straight line, but the line will be tilted to left or right, as shown in Fig. 9–48e or f.

9–27. CHECKING POTENTIOMETERS

An oscilloscope can be used to check the noise (both static and dynamic) of a potentiometer, rheostat, variable resistor, or slider resistance. Static potentiometer "noise" is a result of any current variation due to poor contact, when the contact arm is at rest. Dynamic noise is the amount of irregular current variation, when the contact arm is in motion.

As shown in Fig. 9–49, a constant direct current is applied through the

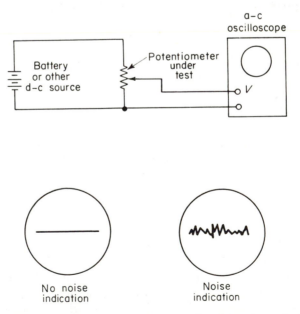

FIGURE 9-49 Checking potentiometers for dynamic and static noise.

potentiometer by means of an external source. A battery is the best source, since it is free of any noise or ripple. An output voltage from the potentiometer is applied to the oscilloscope vertical channel. The internal recurrent sweep can be used, provided that the sweep frequency is above about 100 Hz. If the potentiometer is "quiet," there will be a straight horizontal trace, with no vertical deflection. Any vertical deflection indicates noise.

1. Connect the equipment as shown in Fig. 9–49.
2. Place the oscilloscope in operation (Chapter 5). Switch on the internal recurrent sweep. Set the sync selector to external so that the sweep will not be triggered by noise, if any.

NOTE

An a-c oscilloscope is recommended for this test since the voltage divider action of the potentiometer would move the trace vertically on a d-c oscilloscope.

3. Measure the static noise level, if any, on the voltage-calibrated vertical scale. It is possible that a noise indication could be caused by pickup in the lead wires. If in doubt, disconnect the leads from the potentiometer, but not from the oscilloscope. If the noise is still present, it is pickup noise. If the noise is removed, it is static noise (probably due to poor contact of the potentiometer arm).
4. Vary the potentiometer contact arm from one extreme to the other. Measure the dynamic noise level, if any, on the voltage-calibrated scale. Dynamic noise should not be difficult to distinguish since it occurs only when the contact arm is in motion. (On commercial test units, the contact arm is driven by a motor.)

NOTE

The dynamic noise level will usually be increased if the battery voltage (or other source) is increased. Do not exceed the maximum rated voltage of the potentiometer when making these tests.

9–28. CHECKING RELAYS

An oscilloscope can be used to check the make and break of relay contacts. The presence of contact "bounce," as well as the actual make time and break time of the contacts can be displayed and measured. To be effective, the oscilloscope should be capable of single-sweep operation (Chapter 2). Also, because of the instantaneous nature of the trace, the display should be photographed (unless a storage-type oscilloscope is used).

NOTE

A storage oscilloscope is often used to display instantaneous "one-shot" traces. The storage oscilloscope will hold a display for an indefinite period until it is removed by operating an "erase" control. Because the storage oscilloscope is a special-purpose instrument, it is not discussed in this book. The techniques for operating a storage oscilloscope and for photographing the stored display are covered thoroughly in the instruction manuals for the instruments.

Figure 9–50a is the connection diagram for testing a d-c relay. When S_1 is closed, a d-c voltage from the external battery is applied to the sweep trigger input to initiate a single sweep. Current is also applied through the relay under test, and causes the relay contacts to close. When the contacts close, the positive voltage across R_1 is applied to the vertical channel. When S_1 is opened, the d-c voltage is removed and the relay contacts open, removing the positive voltage from the vertical input. Normally, the opening and closing of the relay contacts produces a rectangular trace similar to that of Fig. 9–50c. If the contacts are bouncing, the display will be similar to that of Fig. 9–50d. The make time and break time of the relay can be measured on the time-calibrated horizontal axis (Fig. 9–50e). The value of R_1 is chosen to limit the relay contact current to the value specified by the manufacturer's data.

Figure 9–50b is the connection diagram for testing an a-c relay. The relay coil is supplied as through the A contact of S_1; dc is applied to the contacts through the B contacts. When S_{1A} is closed, the a-c voltage actuates the relay and closes the contacts. Simultaneously, when S_{1B} is closed, a d-c voltage from the external battery is applied to the sweep trigger input to initiate a single sweep. When the contacts close, the positive voltage across R_1 is applied to the vertical channel. When S_{1A} is opened, the ac is removed from the relay coil, and the contacts open, removing the positive d-c voltage from the vertical input. Normally, the opening and closing of the relay contacts produce a rectangular trace similar to that of Fig. 9–50c. If the contacts are bouncing, the display will be similar to that of Fig. 9–50d. A-c relays may also exhibit contact "chatter." This shows up as a ripple along the normally flat top of the trace. The make time and break time of the relay can be measured along the time calibrated horizontal axis. Again, the value of R_1 is chosen so that the relay contact current limit will not be exceeded.

1.   Connect the equipment as shown in Fig. 9–50a or b, as applicable.
2.   Place the oscilloscope in operation (Chapter 5). Set the oscilloscope to the single-sweep mode. Set the sync selector to external, or as

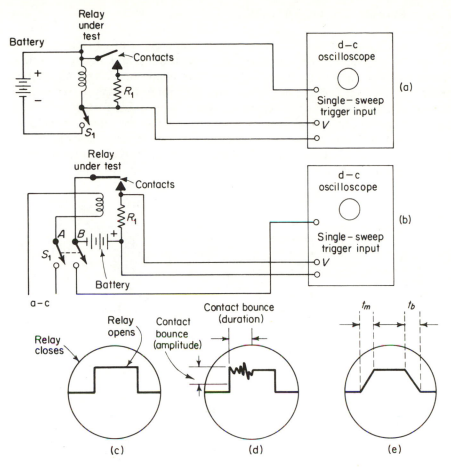

FIGURE 9-50 Checking relays for make time, break time, and contact bounce.

necessary so that the oscilloscope will be triggered by the external d-c voltage. Set up the oscilloscope camera as necessary.

3. If the trace is to be photographed, hold the camera shutter open, close and open switch S_1, then close the camera shutter and develop the picture.

4. Using the developed photo, measure the bounce (if any) amplitude along the vertical axis, and the bounce duration along the horizontal axis (Fig. 9-50d).

5. Measure the make time, t_m (interval between application of voltage and actual closure of contacts), and break time, t_b (interval between interruption of voltage and opening of the relay contacts), along the horizontal axis (Fig. 9-50e).

9-29. CHECKING VIBRATORS

An oscilloscope can be used to check the make and break of vibrator contacts, as well as to display the operation of synchronous and nonsynchronous vibrator power supplies under actual operating condition. Vibrators are best tested by observing contact operation with the vibrator in the power supply. The approximate square wave across the power supply transformer primary is displayed on the oscilloscope, and is compared against an "ideal" waveform.

Figure 9–51a is the connection diagram for testing synchronous and nonsynchronous vibrators. Figure 9–51b is an "ideal" waveform for a typical vibrator power supply.

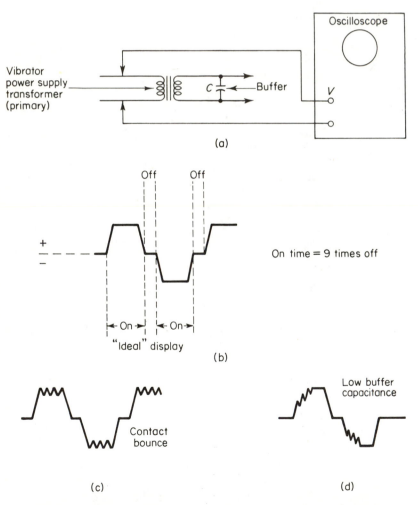

FIGURE 9-51 Checking synchronous and nonsynchronous vibrators and vibrator power supplies.

1. Connect the equipment as shown in Fig. 9–51a.

2. Place the oscilloscope in operation (Chapter 5). Switch on the internal recurrent sweep. Set the sync selector and sweep selector to internal. Adjust the sweep frequency and sync controls for one (or preferably, two) stationary cycles on the oscilloscope screen.

3. Compare the actual oscilloscope display with the "ideal" display of Fig. 9–51b. Measure the display amplitude along the vertical axis. Measure the on-time and off-time of the display along the horizontal axis.

4. Usually, the display amplitudes + and − should be equal, although the actual amplitude is not critical.

5. In most vibrator power supplies, the on-time total should be approximately nine times the off-time total. (The on-time intervals represent the length of time the vibrator contacts are actually closed, and delivering current. The off-time intervals represent the length of time the vibrator contacts are open.)

6. In addition to showing the power supply efficiency (on-time to off-time percentage), the waveform display can also show such conditions as contact bounce (Fig. 9–51c), or insufficient buffer capacitance (Fig. 9–51d).

9–30. CHECKING CHOPPERS

An oscilloscope can be used to check the operation of electromechanical and electronic choppers. Such choppers are similar in function to vibrators, in that they convert direct current into alternating current for further amplification. (A typical example of chopper use is to convert d-c signals into ac for application to an a-c amplifier within an oscilloscope or electronic voltmeter.) Electromechanical choppers differ from vibrators in that they are driven by an external alternating current source, independent of the direct current source to be converted. The resultant output is a square wave that is proportional to the d-c input. Electronic choppers are essentially electronic switches that produce a square-wave output, proportional to d-c input.

Figure 9–52a is the connection diagram for testing an electromechanical chopper.

Figure 9–52b is the connection diagram for testing an electronic chopper.

1. Connect the equipment as shown in Fig. 9–52a or b, as applicable.

2. Place the oscilloscope in operation (Chapter 5). Switch on the internal recurrent sweep. Set the sync selector and sweep selector to internal. Adjust the sweep frequency and sync controls for one (or, preferably, two) stationary cycles on the oscilloscope screen.

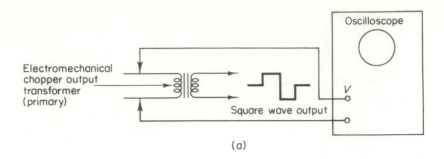

(a)

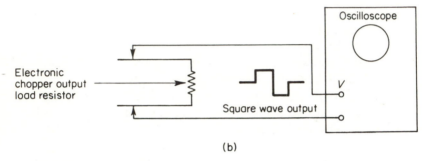

(b)

FIGURE 9-52 Checking electromechanical and electronic choppers.

3. Measure the display amplitude along the vertical axis. Measure the on-time and off-time of the display along the horizontal axis.

NOTE

The square-wave characteristics for choppers are similar to those of vibrators (Fig. 9-51).

4. It is also possible to check the chopper output for electrical noise. All test connections remain the same, except that the d-c input voltage is removed. Under these conditions, any vertical deflection is the result of electrical noise.

NOTE

An electronic chopper should be tested for noise with the d-c input terminals open. An electromechanical chopper should be tested both ways, first with the d-c input terminals shorted, then open.

10

Checking Amplifiers and Amplifier Circuits

An oscilloscope is the most logical instrument for checking amplifiers, whether they are complete audio amplifier systems or the audio circuits of a receiver. The oscilloscope will duplicate every function of an electronic voltmeter in troubleshooting, signal tracing, and performance-testing audio equipment. In addition, the oscilloscope offers the advantage of a visual display for such common audio equipment conditions as distortion, hum, noise, ripple, and oscillation.

This chapter describes the basic procedures for using an oscilloscope to test and troubleshoot audio equipment. A competent technician can expand these procedures to perform a very large number of functions when it is realized that the oscilloscope is basically an electronic voltmeter and/or signal tracer that provides a simultaneous display of audio waveforms.

Although it is possible to use almost any oscilloscope effectively in audio work, it is recommended that the vertical amplifier have a good frequency response up to about 500 kHz. This will ensure that any harmonics or overtones will be properly amplified and displayed, even though the audio range extends to only about 20 kHz, and manufacturers of the most advanced high-fidelity equipment claim band-pass characteristics no higher than 100 kHz.

10-1. AUDIO SIGNAL TRACING WITH AN OSCILLOSCOPE

An oscilloscope is used in a manner similar to that of an electronic voltmeter when signal tracing audio circuits. A sine wave (or square wave) is introduced into the input by means of an external generator. The amplitude and waveform of the input signal are measured on the oscilloscope. In this application, the voltage is measured on the vertical scale as described in Chapter 6. The oscilloscope probe is then moved to the input and output of each stage, in turn, until the final output (usually at a loudspeaker or output transformer) is reached. The gain of each stage is measured as a voltage on the oscilloscope vertical scale. In addition, it is possible to observe any change in waveform from that applied to the input by the external generator. Thus, stage gain and distortion (if any) are established quickly with an oscilloscope.

1. Connect the equipment as shown in Fig. 10–1.
2. Place the oscilloscope in operation (Chapter 5). Switch on the inter-

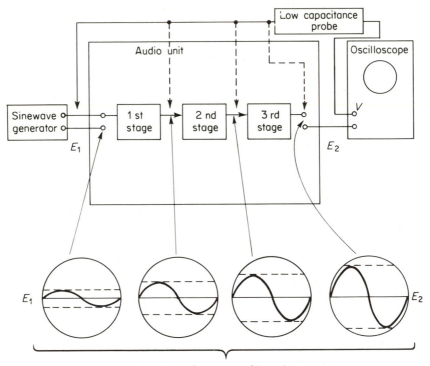

Amplitude increases with each stage
waveform remains substantially the same

FIGURE 10-1   Basic audio signal tracing with an oscilloscope.

nal recurrent sweep. Set the sweep selector and sync selector to internal.

3. Place the generator in operation as described in its instruction manual. Unless otherwise specified by amplifier data, set the generator output frequency to 1000 Hz. Set the generator output level to the value recommended in the amplifier manufacturer's data. Do not overload the amplifier.

4. With the oscilloscope probe (low-capacitance) connected to the generator output (amplifier input), adjust the sweep frequency controls to display one or two complete cycles on the screen.

5. Move the oscilloscope probe to the output of the first amplifier stage. Measure the voltage gain as described in Chapter 6. Compare the waveform at the stage output with that of the stage input. Normally, the waveform should be substantially the same, except for amplitude.

6. Repeat step 5 for each stage of the amplifier, from input to output. If the waveform at a stage output is absent, does not show sufficient gain, or is distorted in any way, it is likely that the particular stage is defective. Usually, voltage/resistance checks or component replacement, will reveal the fault.

NOTE

One factor often overlooked in testing amplifiers is setting the amplifier amplitude and tone controls to their normal operating point, or to some particular point specified in the manufacturer's test data.

7. If it is desired to convert the voltage gain of one stage, a group of stages, or the complete amplifier into decibels, use the equation

$$\text{db gain} = 20 \log \frac{E_2}{E_1}$$

where E_2 = output voltage (of the stage, stages, or complete amplifier)

  E_1 = input voltage

NOTE

Voltage gain is normally measured with a sine-wave input rather than a square wave or pulse.

8.   The gain (or loss) of an audio component, such as a transformer or audio filter, can also be measured as described in steps 5 through 7.

10-2.   CHECKING AUDIO FREQUENCY RESPONSE WITH AN OSCILLOSCOPE

An oscilloscope can be used to obtain a frequency-response curve of an audio amplifier or circuit. In this application, the oscilloscope is used as an audio-frequency voltmeter. The basic method is monitoring the amplifier output with the oscilloscope while applying a constant-amplitude audio signal. The audio signal is varied in frequency (but not amplitude) across the entire audio range. Usually, this is from about 20 Hz to 20 kHz, although some manufacturers specify a response up to 100 kHz. The voltage output at various frequencies across the range is plotted on a graph similar to that shown in Fig. 10–2.

1.   Connect the equipment as shown in Fig. 10–2.
2.   Place the oscilloscope in operation (Chapter 5). Switch on the inter-

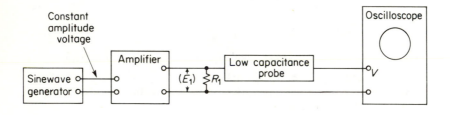

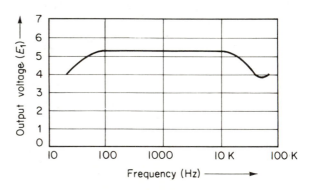

FIGURE 10-2 Checking audio-frequency response with an oscilloscope.

nal recurrent sweep. Set the sweep selector and sync selector to internal.

3. Place the generator in operation as described in its instruction manual. Set the generator output frequency to the lowest point specified in the manufacturer's data. Use a low frequency of approximately 20 Hz in the absence of a specified low limit.

4. With the oscilloscope probe (low-capacitance) connected to the generator output (amplifier input), adjust the sweep frequency controls to display a few cycles on the screen.

5. Set the generator output level to the value recommended in the amplifier manufacturer's data. If no data are available, set the generator output to an arbitrary value. A simple method of determining a satisfactory input level is to monitor the amplifier output with the oscilloscope and increase the generator output until the waveform just starts to flatten, indicating that the amplifier is being overdriven. Then reduce the generator output until the waveform shows no distortion or flattening. When this point is reached, return the oscilloscope probe to the generator output (amplifier input) and measure the voltage. Keep the generator at this voltage throughout the test.

NOTE

Set the amplifier amplitude and tone controls to their normal operating point, or at the particular setting specified in the manufacturer's test data.

6. Record the amplifier output voltage indication on the graph.

7. Without changing the generator output amplitude, increase the generator frequency by 100 Hz, or as specified in the manufacturer's data. Record the new amplifier output voltage indication on the graph. Repeat this process, checking and recording the amplifier output voltage indication at each frequency point throughout the entire audio range. Draw a line on graph paper through each of the check points in order to obtain a frequency response curve. Usually the curve will resemble that of Fig. 10–2, with a flat portion across the center and a roll-off at each end. Some amplifiers and amplifier circuits are designed to provide a high-frequency boost (where the high end of curve increases in amplitude) or low-frequency boost (where the low end shows an amplitude increase). The manufacturer's data must be consulted for this information.

NOTE

Generator output may vary with changes in frequency, a fact often overlooked in making a frequency response test of an amplifier. Even precision laboratory generators can vary in output with changes in frequency. This could result in considerable error. Therefore, it is recommended that the generator output be monitored with the oscilloscope after every change in frequency. Then, if necessary, the generator output amplitude can be reset to the correct value established in step 5. Within extremes, it is more important that the generator output amplitude remain *constant* rather than at some specific value, when making a frequency-response check.

8. Repeat the frequency-response check with the tone control set at each of their positions, or as specified in the manufacturer's data.

NOTE

The load resistor R_1 of Fig. 10–2 is used for power amplifiers. Usually, manufacturers recommend that power amplifiers not be operated without a load. Also, the load serves to stabilize the amplifier output during test. The value of the load resistor should equal the normal output impedance of the amplifier. If practical, the response curve can be run with the amplifier connected to its normal output load (loudspeakers, etc.), thus eliminating the need for a load resistor.

10–3. MEASURING POWER OUTPUT OF AN AMPLIFIER

An oscilloscope can be used as an audio-frequency voltmeter to measure the power output of an amplifier or amplifier circuit. The test connections are identical with that of Fig. 10–2.

1. Connect the equipment as shown in Fig. 10–2.
2. Place the oscilloscope in operation (Chapter 5). Switch on the internal recurrent sweep. Set the sweep selector and sync selector to internal.
3. Place the generator in operation as described in its instruction manual. Set the generator output frequency to the point specified in the manufacturer's data. Use a frequency of 1000 Hz in the absence of a specified value.

4. Set the amplifier gain control to maximum and the tone controls to their normal position, unless otherwise specified in the manufacturer's data.

5. With the oscilloscope probe (low-capacitance) connected to the amplifier output, adjust the sweep frequency controls to display a few cycles on the screen.

6. Set the generator output level to the value recommended in the amplifier manufacturer's data. If no data are available, increase the generator output until the waveform just starts to flatten, indicating that the amplifier is being overdriven. Then reduce the generator output until the waveform shows no distortion or flattening.

7. Measure the maximum output voltage and calculate the power output using the equation

$$P = \frac{E^2}{R}$$

where P = power output

E = maximum voltage indicated on oscilloscope

R = value of load resistor R_1

NOTE

Usually, the power output of an amplifier is based on the rms voltage across a given load, whereas most oscilloscope voltage measurements are peak to peak. Refer to Chapter 6 (Fig. 6–2) for the procedure to convert peak-to-peak voltage to rms voltage.

10–3–1. *Measuring Power Gain of an Amplifier*

To find power gain of an amplifier, it is necessary to find both the input and output power. Input power is found in the same way as output power (the oscilloscope is used to measure voltage at both the input and output), except that the impedance at the input must be known (or calculated). This is not always practical in some circuits, especially where input impedance depends on transistor gain. With input power known (or estimated), the power gain is the ratio of output power to input power.

10–3–2. *Measuring Input Sensitivity of an Amplifier*

An input-sensitivity specification is often used in place of power gain for some audio circuits and amplifiers. Input-sensitivity specifications require a minimum power output with a given voltage input (such as 100-W output with a 1-V input). In such cases, the oscilloscope is used to measure both the input and output voltage. Then the output power is calculated (using the output voltage and load resistance) and compared to the input voltage to find input sensitivity.

10-3-3. Measuring Power Bandwidth of an Amplifier

Many audio circuit or amplifier specifications include a power–band-width factor. Such specifications require that the audio circuit deliver a given power output across a given frequency range. For example, a certain audio amplifier circuit produces full power output up to 20 kHz (as shown in Fig. 10–3) even though the frequency response is flat up to 100 kHz. That is, voltage (without a load) remains constant up to 100 kHz, while power output (across a normal load) remains constant up to 20 kHz.

10-3-4. Measuring Load Sensitivity of an Amplifier

An audio circuit is sensitive to changes in load. This is especially true of audio power amplifiers but can also be the case with voltage amplifiers. An amplifier produces maximum power when the output impedance of the amplifier circuit is the same load impedance. This is shown by the curve of Fig. 10–4 (the load sensitivity for a typical audio amplifier circuit). If the load is twice the output circuit impedance (ratio of 2.0), the output is reduced to approximately 50%. If the load is 40% of the output impedance (ratio of 0.4), the output power is reduced to approximately 25%. Generally, a power amplifier circuit should be checked for load sensitivity during some stage of design. Such a test often shows defects in design that are not easily found with the usual frequency-response and power-output tests.

The circuit for the load-sensitivity test is the same as for frequency response (Fig. 10–2), except that the load resistance R_1 must be variable. (Never use a wire-wound load resistance. The reactance can result in con-siderable error. If a non-wire-wound variable resistance of sufficient wattage

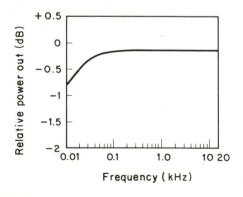

FIGURE 10-3 Typical power band-width graph.

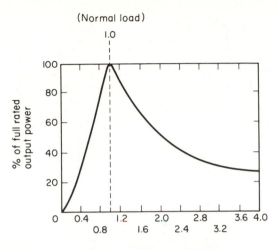

(Normal load)

1.0

FIGURE 10-4 Typical output power versus load impedance (load sensitivity) graph.

is not available, use several fixed carbon or composition resistances arranged to produce the desired resistance values.)

Measure the power output at various load impedance/output ratios. To make a comprehensive test of an audio circuit under design, repeat the load-sensitivity test across the entire frequency range.

10-3-5. Measuring Dynamic Output Impedance of an Amplifier

The load-sensitivity test described in Sec. 10-3-4 can be reversed to find the dynamic output impedance of an audio circuit. The connections and procedures (Fig. 10-2) are the same, except that the load resistance R_1 is varied until maximum output power is found. Power is removed and R_1 is disconnected from the circuit. The d-c resistance of R_1 (measured with an ohmmeter) is equal to the dynamic output impedance. Of course, the values apply only at the frequency of measurement. The test should be repeated across the entire frequency range of the circuit.

10-3-6. Measuring Dynamic Input Impedance of an Amplifier

To find the dynamic input impedance of an audio circuit, use the circuit of Fig. 10-5, or refer to Sec. 8-8. The test conditions should be identical to those for frequency response, power output, and so on. That is, the same audio generator, operating load, oscilloscope, and frequencies should be used.

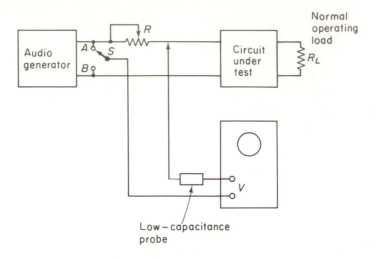

FIGURE 10-5 Measuring the dynamic input impedance of an amplifier.

Adjust the signal source to the frequency (or frequencies) at which the circuit is operated. Move switch S back and forth between positions A and B, while adjusting resistance R until the voltage indication on the oscilloscope is the same in both positions of the switch. Disconnect resistor R from the circuit and measure the d-c resistance of R with an ohmmeter. The d-c resistance of R is equal to the dynamic impedance at the circuit input.

Accuracy of the impedance measurement depends on the accuracy with which the d-c resistance is measured. A noninductive resistance must be used. The impedance found by this method applies only to the frequency used during the test.

10-4. MEASURING AMPLIFIER NOISE AND HUM WITH AN OSCILLOSCOPE

If the vertical channel of an oscilloscope is sufficiently sensitive, the oscilloscope can be used to check and measure the background noise level of an amplifier, as well as to check for the presence of hum, oscillation, and so on. The oscilloscope vertical channel should be capable of a measurable deflection with 1 mV or less, since this is the background noise level of some amplifiers. The basic procedure consists of measuring amplifier output with the gain control at maximum, but without an input signal. The oscilloscope is superior to a voltmeter for noise-level measurement since the frequency and nature of the noise (or other signal) will be displayed visually.

1. Connect the equipment as shown in Fig. 10–6. The load resistor R_1 is used for power amplifiers and should have a value equal to the amplifier's output impedance.

2. Place the oscilloscope in operation (Chapter 5). Switch on the internal recurrent sweep. Set the sweep selector and sync selector to internal.

3. Set the amplifier gain control to maximum and the tone controls to their normal position, unless otherwise specified in the manufacturer's data.

4. Increase the oscilloscope vertical gain control until there is a noise or "hash" indication.

NOTE

It is possible that a noise indication could be caused by pickup in the lead wires. If in doubt, disconnect the leads from the amplifier, but not from the oscilloscope. If the noise is still present, it is pickup noise. If the noise is removed, it is amplifier background noise.

5. Measure the noise voltage. This is the total noise voltage, including hum, background noise, oscillation, and so on.

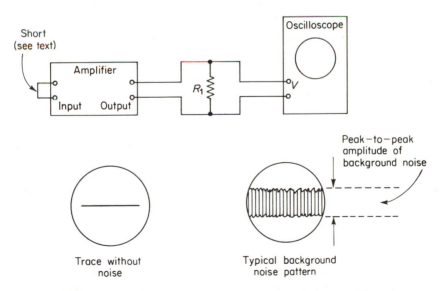

FIGURE 10-6 Measuring amplifier noise and hum with an oscilloscope.

6. If it is suspected that there is line hum present in the amplifier output, set the oscilloscope sync control to line. If a stationary signal pattern appears, this is due to line hum. Measure the amplitude of the line hum, if desired.

7. If a signal appears that is not at the line frequency this can be due to oscillation or stray pickup. Short the amplifier input terminals. If the signal remains, it is probably oscillation. In either case, the oscilloscope can be used to measure both the voltage and frequency of the unknown signal (Chapters 6 and 7).

10-5. CHECKING AMPLIFIER DISTORTION

The major advantage of an oscilloscope over a voltmeter in testing amplifiers is the oscilloscope's ability to display distortion. There are many techniques for checking and measuring distortion in audio equipment. A comprehensive discussion of all methods is beyond the scope of this book, as is a description of all causes and cures for distortion. There are four basic methods that involve the use of an oscilloscope: analysis of sine-wave patterns, analysis of square-wave patterns, measurement of harmonic distortion (fundamental suppression method), and measurement of intermodulation distortion.

10-5-1. Checking Distortion by Sine-wave Analysis

The procedure for checking amplifier distortion by means of sine waves is essentially the same as that described in Sec. 10-1. The primary concern, however, is deviation of the amplifier (or stage) output waveform from the input waveform. If there is no change (except in amplitude), there is no distortion. If there is a change in the waveform, the nature of the change will often reveal the cause of distortion. For example, the presence of second or third harmonics will distort the fundamental as shown in Fig. 10-7.

In practice, analyzing sine waves to pinpoint distortion is a difficult job, requiring considerable experience. Unless the distortion is severe, it may pass unnoticed. Therefore, sine waves are best used where harmonic distortion or intermodulation meters are combined with oscilloscopes for distortion or intermodulation meters are combined with oscilloscopes for distortion analysis. If an oscilloscope is to be used alone, square waves provide the best basis for distortion analysis.

10-5-2. Harmonic Distortion Analysis

No matter what amplifier circuit is used, or how well the circuit is designed, there is always the possibility of odd or even harmonics being present with the fundamental. These harmonics combine with the fundamental and produce distortion, as is the case when any two signals are combined.

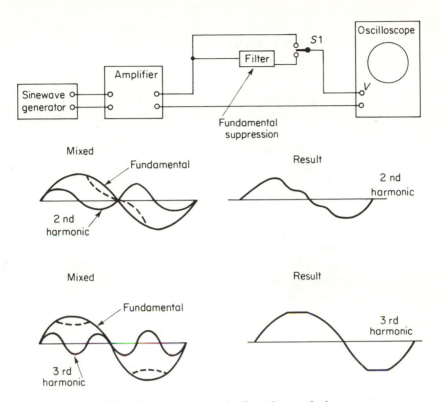

FIGURE 10-7 Harmonic distortion analysis.

Commercial harmonic distortion meters operate on the *fundamental suppression* principle. As shown in Fig. 10–7, a sine wave is applied to the amplifier input, and the output is measured on the oscilloscope. The output is then applied through a filter that suppresses the fundamental frequency. Any output from the filter is then the result of harmonics. This output is also displayed on the oscilloscope where the signal can be checked for frequency to determine the harmonic content. For example, if the input was 1000 Hz, and the output after filtering was 3000 Hz, this would be a result of third harmonic distortion.

The percentage of harmonic distortion can also be determined by this method. For example, if the output without filter was 100 mV, and with filter was 3 mV, this would indicate a 3% harmonic distortion.

1.   Connect the equipment as shown in Fig. 10–7.

NOTE

In some commercial harmonic distortion meters, the filter is tunable, so that the amplifier can be tested over a wide range of

239

fundamental frequencies. In other harmonic distortion meters, the filter is fixed frequency, but can be detuned slightly to produce a sharp null.

2. Place the oscilloscope in operation (Chapter 5). Switch on the internal recurrent sweep. Set the sweep selector and sync selector to internal.

3. Set the amplifier amplitude and tone controls to their normal operating point, or at the particular setting specified in the manufacturer's test data.

4. Place the generator in operation as described in its instruction manual. Set the generator output frequency to the filter null frequency. Set the generator output amplitude to the value recommended in the amplifier manufacturer's data. In the absence of such data, set S_1 to the position without the filter, and increase the generator output until the waveform just starts to flatten indicating that the amplifier is being overdriven. Then reduce the generator output until the waveform shows no distortion or flattening.

5. If necessary, adjust the sweep frequency controls to display a few cycles on the screen.

6. Measure the voltage with switch S_1 in the position without the filter. Record this value as E_1.

7. Set switch S_1 to the position with the filter. Adjust the filter for the deepest null indication on the oscilloscope. Record this value as E_2.

8. Calculate the total harmonic distortion using the equation

$$D = 100 \frac{E_2}{E_1}$$

where D = percentage of total harmonic distortion

E_1 = output before filtering

E_2 = output after filtering

9. If the filter is tunable, select another frequency, tune the generator to that frequency, and repeat the procedure (steps 4–8).

When a circuit or amplifier is tested over a wide frequency range for harmonic distortion, and the results plotted on a graph similar to that of Fig. 10–8, the percentage is known as *total harmonic distortion* (THD). Note that the THD shown in Fig. 10–8 is less than 0.2%. Also note that harmonic distortion can vary with frequency and power output.

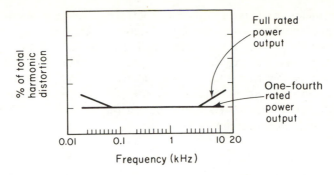

FIGURE 10-8 Typical total harmonic distortion (THD) graph.

10-5-3. Intermodulation Distortion Analysis

When two signals of different frequency are mixed in an amplifier, there is a possibility of the lower-frequency signal amplitude modulating the higher-frequency signal. This produces a form of distortion, known as *intermodulation distortion*.

Commercial intermodulation distortion meters consist of a signal generator and high-pass filter as shown in Fig. 10–9. The signal generator por-

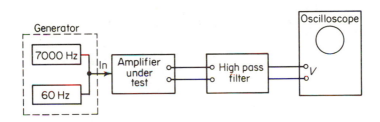

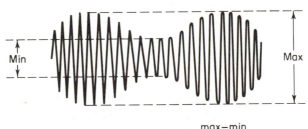

$$\text{Intermodulation (\%)} = 100 \times \frac{\text{max} - \text{min}}{\text{max} + \text{min}}$$

FIGURE 10-9 Intermodulation distortion analysis.

tion of the meter produces a high-frequency signal (usually about 7000 Hz) which is modulated by a low-frequency signal (usually 60 Hz). The mixed signals are applied to the amplifier input. The amplifier output is connected through a high-pass filter to the oscilloscope vertical channel. The high-pass filter removes the low-frequency (60-Hz) signal. Therefore, the only signal appearing on the oscilloscope vertical channel should be the high-frequency (7000-Hz) signal. If any 60 Hz is present on the display, it is being passed through as modulation on the 7000-Hz signal.

1. Connect the equipment as shown in Fig. 10–9.
2. Place the oscilloscope in operation (Chapter 5). Switch on the internal recurrent sweep. Set the sweep selector and sync selector to internal.
3. Set the amplifier amplitude and tone controls to their normal operating point, or at the particular setting specified in the manufacturer's test data.
4. Place the generator in operation as described in its instruction manual.
5. If necessary, adjust the sweep frequency controls to display a few cycles on the screen.
6. Measure the vertical dimensions MAX and MIN (Fig. 10–9) in screen divisions.
7. Calculate the total intermodulation distortion using the equation

$$IM(\%) = 100 \left(\frac{MAX - MIN}{MAX + MIN} \right)$$

where $IM(\%)$ = percentage of total intermodulation

MIN = modulation minimum dimension

MAX = modulation maximum dimension

8. If desired, repeat the intermodulation measurement at various settings of the amplifier gain and tone controls.

Figure 10–10 shows an intermodulation test circuit that can be fabricated in the shop or laboratory. Note that the high-pass filter is designed to pass signals above about 200 Hz. The purpose of the 39- and 10-kΩ resistors is to set the 60-Hz signals at four times the 7-kHz signal. Most audio generators provide for a line-frequency output (60 Hz) that can be used as the low-frequency modulation source.

When using the circuit of Fig. 10–10, set the generator line-frequency (60-Hz) output to some fixed value (1 V, 2 V, etc.). Then set the generator

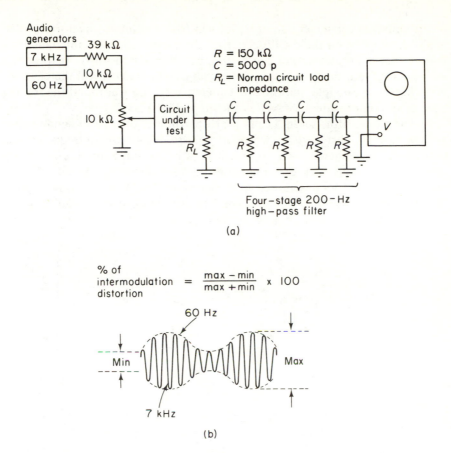

(a)

$$\% \text{ of intermodulation distortion} = \frac{max - min}{max + min} \times 100$$

(b)

FIGURE 10-10 Measuring percentage of intermodulation distortion. $R = 150$ RΩ; $c = 5000$ pf; R_L = normal circuit load impedance.

audio output (7 kHz) to the same value. If the line-frequency output is not adjustable, measure the actual value of the line-frequency output, and then set the generator audio output to the same value.

10-5-4. Checking Distortion by Square-Wave Analysis

The procedure for checking distortion by means of square waves is essentially the same as for sine waves, as described in Sec. 10-5-1. Distortion analysis is more effective with square waves because of their high odd-harmonic content and because it is easier to see a deviation from a straight line with sharp corners, than from a curving line. As in the case of sine-wave distortion testing, square waves are introduced into the amplifier input, while the output is monitored on an oscilloscope. The primary concern is deviation of the amplifier (or stage) output waveform from the input waveform (which is also monitored on the oscilloscope). If the oscilloscope has the dual-trace feature, the input and output can be monitored simultaneously. If there is a change in waveform, the nature of the change will often reveal the cause of distortion. For example, poor high-frequency response will round the trailing edge of the output square wave, as shown in Fig. 10–11.

The third, fifth, seventh, and ninth harmonics of a clean square wave are emphasized. Therefore, if an amplifier passes a given audio frequency and produces a clean square-wave output, it is safe to assume that the frequency response is good up to at least nine times the fundamental frequency. For example, if an amplifier passes a clean square wave at 3000 Hz, it shows a good response up to 27 kHz, which is beyond the top limit of the audio range.

A square wave provides a controlled signal to evaluate amplifier perfor-

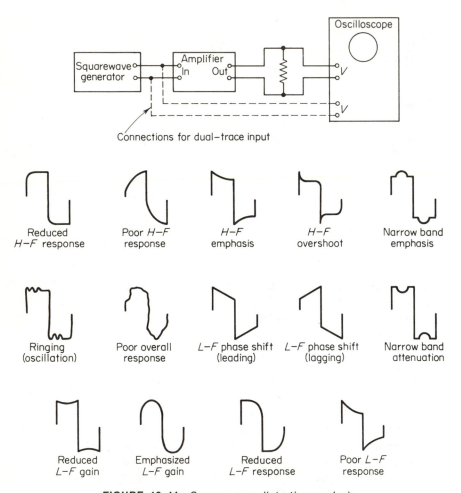

FIGURE 10-11 Square-wave distortion analysis.

mance in response to simultaneous signals of many frequencies (harmonics of the square wave), which is what the amplifier sees when amplifying complex waveforms of musical instruments or voices. However, it should be noted that the actual response check of an amplifier should be made using a sine-wave signal. This is especially important in limited-bandwidth amplifiers (voice amplifiers). The square-wave signal provides a quick check on amplifier performance and gives an estimate of overall amplifier efficiency not readily apparent when using a sine-wave signal. Whether a sine wave or square wave is used for testing the amplifier, it important that the manufacturer's specifications be known in order to make a better judgment of performance.

The square-wave output of the signal generator must be extremely flat so that the signal does not contribute to any distortion that may be observed when evaluating amplifier response. A d-c oscilloscope introduces the least distortion when testing with square waves. Generally, when checking amplifier response, the frequency of the square-wave input should be varied from the low end of the amplifier band-pass up toward the upper end. However, because of the harmonic content of square waves, distortion occurs before the upper end of the amplifier band-pass is reached.

1. Connect the equipment as shown in Fig. 10–11. A load resistor must be used for power amplifiers and should have a value equal to the amplifier's output impedance.

2. Place the oscilloscope in operation (Chapter 5). Switch on the internal recurrent sweep. Set the sweep selector and sync selector to internal.

3. Set the amplifier amplitude and tone controls to their normal operating point, or at the particular setting specified in the manufacturer's test data.

4. Place the generator in operation as described in its instruction manual. Set the generator output frequency to 1000 Hz or as specified in the amplifier manufacturer's test data. If no data are available, increase the generator output until the waveform no longer increases in amplitude and/or shows distortion. Then reduce the generator output until the waveform shows no distortion.

5. If necessary, adjust the sweep frequency controls to display one (or possibly two) cycles on the screen.

6. Compare the amplifier output waveform with the input (generator output) waveform. If the output is identical to the input, except possibly for amplitude, there is no distortion. If the output is not identical, compare it with the typical response patterns of Fig. 10–11.

7. If desired, repeat the square-wave distortion analysis at other settings of the amplifier gain and tone controls, as well as at other generator frequencies.

10-5-5. Measuring Distortion with a Dual-Trace Oscilloscope

A dual-trace oscilloscope with an ADD function (both channels added algebraically) can be used to measure distortion of an amplifier. This type of measurement is especially valuable when the slope of a waveform must be faithfully reproduced by the amplifier. Figure 10–12 shows the testing of an amplifier circuit using a triangular wave. The same connections and procedures can be used with any type of waveform. The test is most effective when the input waveform duplicates that normally used in the circuit.

1. Connect the equipment as shown in Fig. 10–12.
2. Place the oscilloscope in operation (Chapter 5).
3. Connect the channel A probe to the input and the channel B probe to the output, as shown.
4. Adjust the gain of the two channels so that both traces are the same height. Usually, this means that the channel B (output) gain be set much lower than the channel A (input) gain.
5. Adjust the position controls as necessary to superimpose the waveforms on each other.

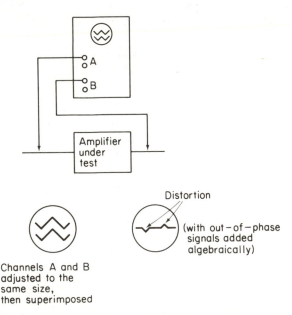

Channels A and B
adjusted to the
same size,
then superimposed

Distortion

(with out-of-phase
signals added
algebraically)

FIGURE 10–12 Measuring distortion using a dual-trace oscilloscope with an ADD function.

6. Set the input selector to measure ac. Operate the controls as necessary to select the ADD function.

7. If the channel A and channel B signals are normally in-phase, invert one channel (channel B) by operating the oscilloscope invert control. Do not use the invert function if channel A and channel B are normally out-of-phase.

8. Adjust the fine vertical sensitivity (if the oscilloscope has such a control) to minimize any remaining waveform.

9. Any waveform that remains (with out-of-phase signals added algebraically) equals distortion. If the two waveforms are exactly the same amplitude, and there is no distortion, the waveforms cancel and a straight horizontal line remains on the screen.

10-6. MEASURING AMPLIFIER PHASE SHIFT

An amplifier can be checked for phase shift between input and output by either of the methods described in Chapter 7 (Secs. 7–8 and 7–9). The test connection diagrams are shown in Fig. 10–13.

10-7. MEASURING AMPLIFIER SLEW RATE

The slew rate of an amplifier is the maximum rate of change of the output voltage, with respect to time, that the op amp is capable of producing while maintaining linear characteristics (symmetrical output without clipping). Slew rate is expressed in terms of

$$\frac{\text{difference in output voltage}}{\text{difference in time}} \quad \text{or} \quad \frac{dV_o}{dt}$$

Usually, slew rate is listed in terms of *volts per microsecond*. For example, if the output voltage from an amplifier is capable of changing 7 V in 1 μs, the slew rate is 7. Generally, the coupling and bypass capacitors used in an amplifier have the most effect on slew rate. At higher frequencies, the current required to charge and discharge a capacitor can limit available current to succeeding stages or load, and thus result in lower slew rates.

The major effect of slew rate on amplifier performance is on output power. All other factors being equal, a lower slew rate results in lower power output. Slew rate and the term *full power response* of an amplifier are directly related. Full power response is the maximum frequency for which rated out-

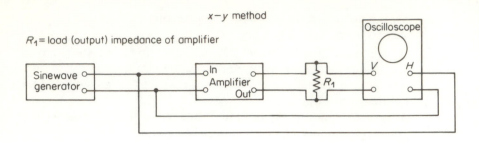

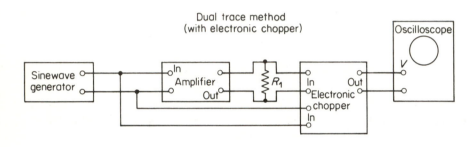

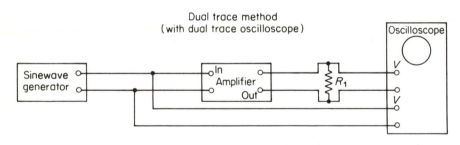

FIGURE 10-13 Measuring amplifier phase shift.

put voltage can be obtained for a sine-wave signal, with a specified load, and without distortion due to slew-rate limiting.

The slew rate versus full-power response relationship can be shown as

$$\text{slew rate (V/s)} = 6.28 \times F_M \times E_O$$

where F_M is the full-power response frequency in Hz, and E_O is the peak output voltage (one-half the peak-to-peak voltage). For example, if E_O is 13 V (one half a peak-to-peak voltage of 26 V) at a frequency of 30 kHz, the slew rate is

$$6.28 \times 30,000 \times 13 \approx 2,449,200 \text{ V/s} \approx 2.45 \text{ V/}\mu s$$

The equation can be turned around to find the full-power response frequency. For example, assume that an amplifier is rated as having a slew rate of 2.5 V/μs and a peak-to-peak output of 20 V (E_O = 10 V). Find the full-power response frequency F_M as follows:

$$F_M = \frac{2.5 \text{ V/}\mu s}{6.28 \times 10} = \frac{2,500,000}{62.8} \approx 40,000 \text{ Hz} \approx 40 \text{ kHz}$$

With a constant output load, the power output of an amplifier depends on output voltage. In turn, all other factors being equal, output voltage depends on slew rate. The graph of Fig. 10–14 shows the relationship among slew rate, full-power frequency response, and output voltage. For example, if slew rate is 6 V/μs, the maximum output voltage (peak to peak) is about 20 V at a frequency of 100 kHz, and vice versa.

An oscilloscope can be used to observe and measure the slew rate of an amplifier. The basic procedure is to measure the slope of the output waveform of a square-wave input signal, as shown in Fig. 10–15. The input square wave

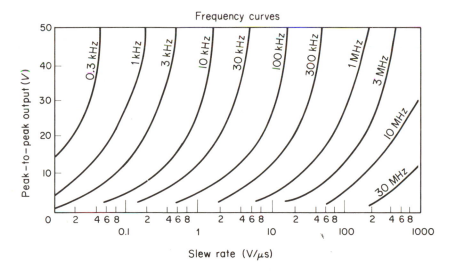

FIGURE 10–14 Typical slew-rate graphs.

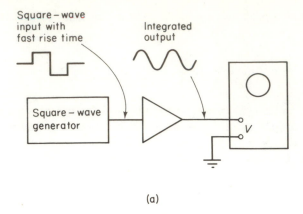

(a)

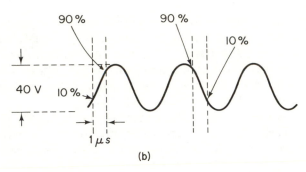

(b)

FIGURE 10-15 Example of slew-rate measurement.

must have a rise time that exceeds the slew-rate capability of the amplifier. Thus, the output appears on the oscilloscope screen as an integrated wave, not as a square wave. In the example shown, the output voltage rises (and falls) about 40 V in 1 μs. Referring back to Fig. 10–14, a slew rate of 40 (40 V/μs) means that the amplifier should be capable of delivering a 40-V peak-to-peak output with full power at frequencies up to about 300 kHz.

11

Checking Communications Equipment

The role of an oscilloscope in testing communications equipment is often overlooked. The TV bench service technician finds the oscilloscope almost indispensable, but the technician who services communications receivers tends to avoid the oscilloscope. When, however, it is realized that an oscilloscope can provide an effective instrument for AM front end and IF alignment, FM front end, IF, and detector alignment, as well as a modulation meter for testing transmitters, the communications service technician can put the oscilloscope in its true perspective.

11-1. CHECKING MICROPHONES

A microphone can be checked for distortion and frequency response using an oscilloscope. Distortion is checked by applying a sine wave to a loudspeaker placed near the microphone, and monitoring the microphone output on an oscilloscope. Frequency response is checked in essentially the same way, by varying the sine wave over the desired test range of the microphone.

1. Connect the equipment as shown in Fig. 11–1.

2. Place the oscilloscope in operation (Chapter 5). Switch on the internal recurrent sweep. Set the sweep selector and sync selector to internal.

3. Place the generator and amplifier (if any) in operation as described in their instruction manuals. Set the generator output frequency to the low limit of the microphone. Adjust the generator and amplifier controls for a suitable pattern on the oscilloscope.

4. Check the oscilloscope pattern for distortion. If there is doubt as to the origin of any observed distortion, temporarily connect the generator output to the amplifier input. If distortion is removed, the cause is in the microphone (assuming that the loudspeaker is distortion-free).

NOTE

For the most accurate results, the microphone should be shielded from all sound sources, except the loudspeaker.

5. Without changing the generator output amplitude, vary the generator frequency over the entire test range of the microphone. Check for any change in amplitude on the oscilloscope pattern.

NOTE

A response curve can be made for the microphone by following the procedures of Sec. 10–2.

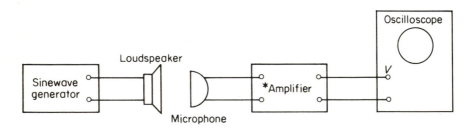

*Amplifier may be omitted
if oscilloscope vertical
gain is sufficient

FIGURE 11-1 Checking microphones with an oscilloscope.

11-2. CHECKING TRANSMITTER AMPLITUDE MODULATION WITH AN OSCILLOSCOPE

The main use for an oscilloscope in communications work is to measure percentage of modulation and uniformity or linearity of modulation. The use of an oscilloscope for modulation checks is not new. There are many variations of the basic technique, each of which is discussed in the following paragraphs of this section.

11-2-1. Direct Measurement of Modulation Envelope with a High-Frequency Oscilloscope

If the vertical channel response of the oscilloscope is capable of handling the transmitter output frequency, the output can be applied through the oscilloscope vertical amplifier. The basic test connections are shown in Fig. 11-2. The procedure is as follows:

1. Connect the oscilloscope to the antenna jack, or the final RF amplifier of the transmitter, as shown in Fig. 11-2. Use one of the three alternatives shown, or the modulation measurement described in the transmitter service literature.

2. Key the transmitter (press the push-to-talk switch) and adjust the oscilloscope controls to produce displays as shown. You can either speak into the microphone (for a rough check of modulation), or you can introduce an audio signal (typically 400 or 1000 Hz) at the microphone jack input (for a precise check of modulation). Note that Fig. 11-2 provides simulations of typical oscilloscope displays during modulation tests.

3. Measure the vertical dimensions shown as *A* and *B* in Fig. 11-2 (the crest amplitude and the trough amplitude). Calculate the percentage of modulation using the equation of Fig. 11-2. For example, if the crest amplitude (*A*) is 63 (63 screen divisions, 6.3 V, and so on) and the trough amplitude (*B*) is 27, the percentage of modulation is

$$\frac{63 - 27}{63 + 27} \times 100 = 40\%$$

Make certain to use the same oscilloscope scale for both crest (*A*) and trough (*B*) measurements. Keep in mind when making modulation

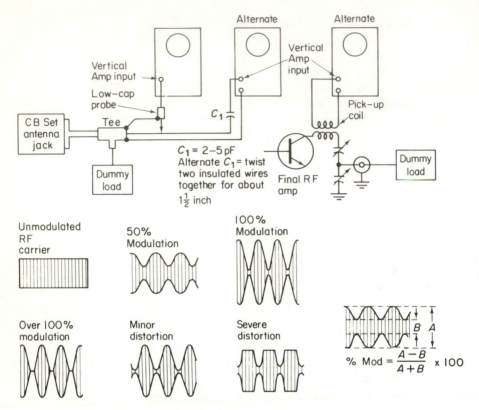

FIGURE 11-2 Direct measurement of modulation envelope with a high-frequency oscilloscope.

measurements, or any measurement that involves the transmitter, that the RF output (antenna connector) must be connected to an antenna or dummy load.

11-2-2. Direct Measurement of Modulation Envelope with a Low-Frequency Oscilloscope

If the oscilloscope is not capable of passing the transmitter frequency, the transmitter output can be applied directly to the vertical deflection plates of the oscilloscope CRT. However, there are two drawbacks to this approach. First, the vertical plates may not be readily accessible. Next, the voltage output of the final RF amplifier may not produce sufficient deflection of the oscilloscope trace.

The test connections and modulation patterns are essentially the same as those shown in Fig. 11–2, and the procedures are the same as those described in Sec. 11–2–1.

11-3. CHECKING TRANSMITTER AMPLITUDE MODULATION WITH TRAPEZOIDAL PATTERNS

The trapezoidal technique has an advantage in that it is easier to measure straight-line dimensions than curving dimensions. Thus, any nonlinearity in modulation may easily be checked with a trapezoid. In the trapezoidal method, the modulated carrier amplitude is plotted as a function of modulating voltage rather than as a function of time. The basic test connections are shown in Fig. 11-3.

1. Connect the oscilloscope to the final RF amplifier and modulator. As shown in Fig. 11-3, use either the capacitor connection or the pickup coil for the RF (oscilloscope vertical input). However, for best results, connect the transmitter output directly to the deflection

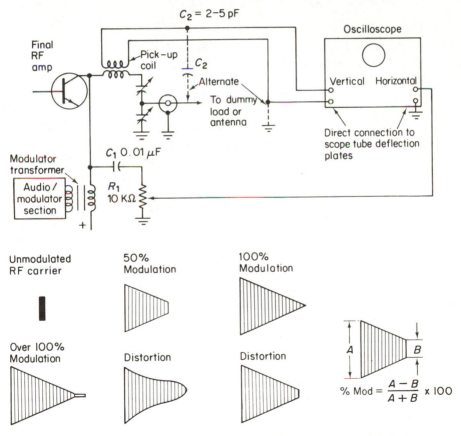

FIGURE 11-3 Trapezoidal measurement of the modulation envelope.

255

plates of the oscilloscope CRT. The oscilloscope amplifiers may be nonlinear and can cause the modulation to appear distorted.

2. Key the transmitter and adjust the controls (oscilloscope controls and R_1) to produce a display as shown.

3. Measure the vertical dimensions shown as A (crest) and B (trough) on Fig. 11–3, and calculate the percentage of modulation using the equation given. For example, if the crest amplitude (A) is 80, and the trough amplitude (B) is 40, using the same scale, the percentage of modulation is

$$\frac{80 - 40}{80 + 40} \times 100 = 33\%$$

Again, before transmitting, make sure that the transmitter output is connected to an antenna or dummy load.

11–4.   DOWN-CONVERSION MEASUREMENT OF THE MODULATION ENVELOPE

If the oscilloscope is not capable of passing the transmitter carrier signals, and the transmitter output is not sufficient to produce a good indication when connected directly to the oscilloscope CRT, it is possible to use a down-converter test setup. One method requires an external RF generator and an IF transformer. The other method uses a receiver capable of monitoring the transmitter frequencies.

The RF generator method of down-conversion is shown in Fig. 11–4. In this method, the RF generator is tuned to a frequency above or below the transmitter frequency by an amount equal to the IF transformer frequency. For example, if the IF transformer is 455 kHz, tune the RF generator to a frequency 455 kHz above (or below) the transmitter frequency.

The receiver method of down-conversion is shown in Fig. 11–5. With this method, the receiver is tuned to the transmitter frequency, and the oscilloscope input signal is taken from the last IF stage output through a 30-pF capacitor.

With either method of down-conversion, the RF generator or receiver is tuned for a maximum indication on the oscilloscope screen. Once a good pattern is obtained, the rest of the procedure is as described in Sec. 11–2–1.

The author does not generally recommend the down-conversion methods, except as a temporary measure. There are a number of relatively inexpensive oscilloscopes available that will pass signals up to and beyond the 50-MHz range.

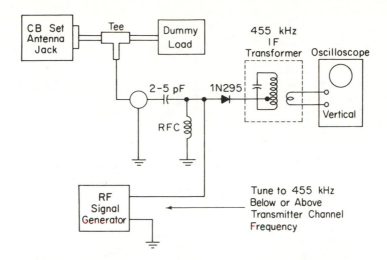

FIGURE 11-4 Down-conversion method of modulation measurement using a 455-kHz IF transformer.

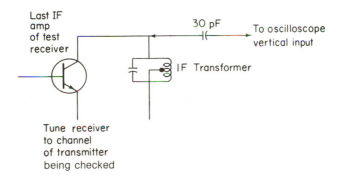

FIGURE 11-5 Down-conversion method of modulation measurement using a receiver.

11-5. LINEAR DETECTOR MEASUREMENT OF THE MODULATOR ENVELOPE

If you must use an oscilloscope that will not pass the carrier frequency of the transmitter, you can use a linear detector. However, the oscilloscope must have a d-c input. The basic test connections for linear detection of the modulation envelope are shown in Fig. 11-6. The basic test procedure is as follows:

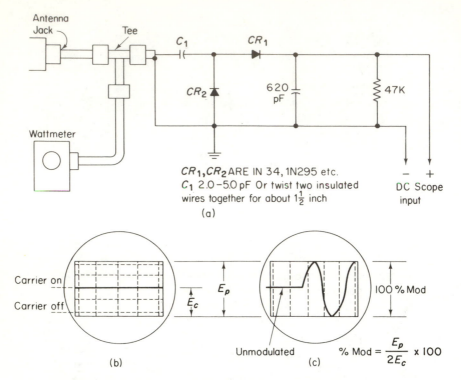

CR_1, CR_2 ARE IN 34, 1N295 etc.
C_1 2.0–5.0 pF Or twist two insulated
wires together for about $1\frac{1}{2}$ inch

(a)

$$\% \text{ Mod} = \frac{E_p}{2E_c} \times 100$$

(b)   (c)

FIGURE 11-6 Pace modulation detector for measurement of the
modulation envelope with a low-frequency oscilloscope.

1. Connect the transmitter output to the oscilloscope through the linear detector circuit as shown in Fig. 11-6. Make certain to include the dummy load (or wattmeter, as shown).

2. With the transmitter not keyed, adjust the oscilloscope position controls to place the trace on a reference line near the bottom of the screen, as shown in Fig. 11-6b.

3. Key the transmitter, but do not apply modulation. Adjust the oscilloscope gain control to place the top of the trace at the center of the screen, as shown in Fig. 11-6b. It may be necessary to switch the transmitter off and on several times to adjust the trace properly, since the position and gain controls of most oscilloscopes interact.

4. Measure the distance (in scale divisions) of the shift between the carrier (step 3) and no-carrier (step 2) traces. For example, if the screen has a total of 10 vertical divisions, and the no-carrier trace is at the bottom or zero line, there is a shift of five scale divisions to the centerline.

5. Key the transmitter and apply modulation. Do not touch either the position or gain controls of the oscilloscope.

6. Find the percentage of modulation using the equation shown in Fig. 11-6. For example, assume that the shift between the carrier and no-carrier trace is five divisions, and that the modulation produces a peak-to-peak envelope of eight divisions. The percentage of modulation is

$$\frac{8}{2 \times 5} \times 100 = 80\%$$

11-6. MODULATION NOMOGRAM

Figure 11-7 is a nomogram that can be used with direct measurement techniques (Secs. 11-2-1 and 11-2-2) or the trapezoidal technique (Sec. 11-3-5) to find the percentage of modulation. To use Fig. 11-7, measure the values of the crest (or maximum) and trough (or minimum) oscilloscope patterns.

The percentage of modulation is found by extending a straightedge from the measured value of the crest or maximum (given as A on Fig. 11-7) on the scale to the measured value of the trough or minimum (given as B) on the scale. The percentage of modulation is found where the straightedge crosses the diagonal scale. The crest and trough may be measured in any units (volts, vertical scale divisions, etc.), as long as both crest and trough are measured in the same units. The dashed line in Fig. 11-7 is used to illustrate the percentage of modulation example of Sec. 11-2-1.

11-7. CHECKING TRANSMITTER SINGLE-SIDEBAND (SSB) MODULATION WITH AN OSCILLOSCOPE

An oscilloscope can be used to display the carrier of a single-sideband (SSB) modulated radio wave. Best results are obtained when the frequency response of the oscilloscope vertical channel is well above the transmitter frequency.

11-7-1. Direct Measurement of SSB Modulation Envelope with a High-Frequency Oscilloscope

If the vertical channel of the oscilloscope is capable of handling the output frequency of an SSB transmitter (including the sidebands), the output can be applied through the oscilloscope vertical amplifier. The basic test connections are shown in Fig. 11-8. The procedure is as follows:

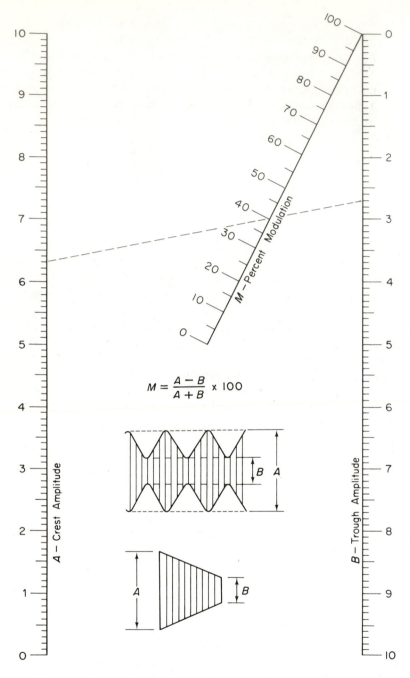

$$M = \frac{A - B}{A + B} \times 100$$

FIGURE 11-7   Pace modulation nomogram.

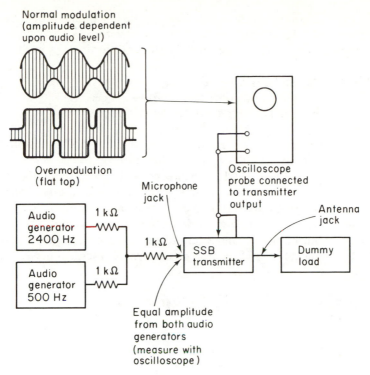

FIGURE 11-8 Checking transmitter single-sideband (SSB) modulation.

1. Connect the oscilloscope to the antenna jack or the final RF amplifier of the transmitter (generally a linear amplifier), as shown in Fig. 11–8.

2. Apply two simultaneous, *equal-amplitude* audio signals for modulation, such as 500 Hz and 2400 Hz. The audio signals must be free of distortion, noise, and transients. The two audio signals *must not* have a direct harmonic relationship such as 500 Hz and 1500 Hz.

3. Check the modulation envelope against the patterns of Fig. 11–8, or against patterns shown in the SSB transmitter service literature. Note that the typical SSB modulation envelope resembles the 100% AM modulation envelope, except that the amplitude of the entire SSB waveform varies with the strength of the audio signal. Thus, the percentage of modulation calculations that apply to AM cannot be applied to SSB.

4. Increase the amplitude of both audio modulating signals, making certain to maintain both signals at equal amplitudes. When peak SSB power output is reached, the modulation envelope will "flat

top'' as shown in Fig. 11–8. That is, the instantaneous RF peaks of the SSB signal reach saturation, even with less than peak audio signal applied. This overmodulated condition results in distortion.

11-7-2. Direct Measurement of SSB Modulation Envelope with a Low-Frequency Oscilloscope

If the oscilloscope is not capable of passing the SSB transmitter frequency, the transmitter output can be applied directly to the vertical deflection plates of the oscilloscope CRT. However, as in the case of AM modulation test, there are two drawbacks to this approach for SSB. First, the vertical plates may not be readily accessible. Next, the voltage output of the final RF amplifier may not produce sufficient deflection of the oscilloscope trace.

The test connections and modulation patterns are essentially the same as those shown in Fig. 11–8, and the procedures are as described in Sec. 11–7–1.

11-8. CHECKING TRANSMITTER MODULATOR CHANNELS WITH AN OSCILLOSCOPE

The modulator channel of a transmitter is essentially an audio amplifier. As such, the modulation channel can be checked for frequency response, power output, noise and hum, distortion, and phase shift as described in Chapter 10. An oscilloscope can also be used as an audio signal tracer for the modulator channel.

11-9. CHECKING TRANSMITTER RF CHANNEL FOR FREQUENCY MULTIPLICATION WITH AN OSCILLOSCOPE

The various multiplier stages of the RF channel in a transmitter must be tuned to different frequencies, which are multiples of the fundamental or oscillator. Although it is possible to tune each multiplier with the aid of a frequency meter, there may be some confusion since most frequency meters respond to harmonics. This confusion of harmonics can be especially difficult when the transmitter is first placed in operation. An oscilloscope can be used to ensure that each stage is at the desired frequency, and not at an undesired harmonic. This is done by displaying the stage signals as Lissajous patterns, with the oscillator or fundamental as the reference signal.

1.   Connect the equipment as shown in Fig. 11–9.

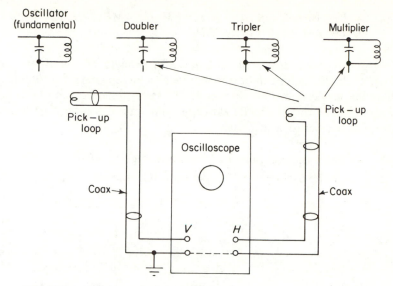

FIGURE 11-9 Checking transmitter RF channel for frequency multiplication with an oscilloscope.

NOTE

Do not use the oscilloscope amplifiers. Make both the horizontal and vertical connections directly to the oscilloscope CRT.

2. Place the oscilloscope in operation (Chapter 5). Switch off the internal recurrent sweep. Set the sync selector to external.

3. Place the transmitter in operation as described in the instruction manual. Set the transmitter for an unmodulated carrier output. Make certain that the transmitter output is connected to a dummy load or to an antenna.

4. In turn, connect the horizontal channel to the tank circuit of each stage to be measured. Leave the vertical channel connected to the oscillator (fundamental) tank.

5. Check whether the stage is passing the fundamental, or multiplying (doubling, tripling, or quadrupling, etc.) the fundamental. The stages can be checked at all multiples using the Lissajous patterns, as described in Sec. 7-4.

6. If necessary, adjust the stage tuning circuit until the desired pattern is obtained and stands still.

11-10. ALIGNMENT OF AM AND FM IF AMPLIFIERS WITH AN OSCILLOSCOPE

The response characteristics of AM and FM receiver IF amplifiers can be checked, or the IF stages can be aligned, using a sweep generator/oscilloscope combination, as described in Chapter 8. The sweep generator must be capable of sweeping over the entire IF range. If maximum accuracy is desired, a marker generator must also be used.

1. Connect the equipment as shown in Fig. 11-10.
2. Place the oscilloscope in operation (Chapter 5). Switch off internal

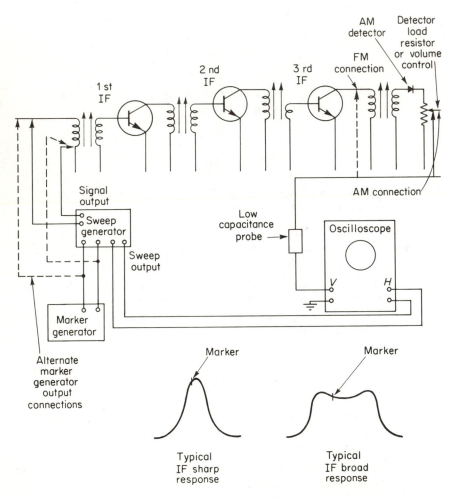

FIGURE 11-10 Alignment of IF amplifiers with an oscilloscope.

recurrent sweep. Set the oscilloscope sweep selector and sync selector to external.

3. Place the sweep generator in operation as described in its instruction manual. Switch sweep generator blanking control on or off as desired. Adjust the sweep generator to cover the complete IF range. Usually, AM IF center frequency is 455 kHz and requires a sweep of about 30 kHz wide; FM IF center frequency is 10.7 MHz and requires a sweep of about 300 kHz.

4. Check the IF response curve appearing on the oscilloscope against those of Fig. 11-10, or against the receiver specifications.

5. If it is desired to determine the exact frequencies at which IF response occurs, the marker generator can be adjusted until the marker pip is aligned at the point of interest. The frequency, or band of frequencies, can be read from the marker generator frequency dial.

6. The amplitude of any point on the response curve can be measured directly on the oscilloscope (assuming that the vertical system is voltage calibrated).

7. Adjust the IF alignment controls to produce the desired response curve, as specified in the receiver service data.

11-11. ALIGNMENT OF AM AND FM FRONT END WITH AN OSCILLOSCOPE

The response characteristics of AM and FM receiver RF stages (RF amplifier, mixer or first detector, oscillator) or "front end" can be checked, or aligned, using a sweep generator/oscilloscope combination, as described in Chapter 8. The procedure is essentially the same as for IF alignment (Sec. 11-10) except that the sweep generator output is connected to the antenna input of the receiver, whereas the input to the first detector or mixer is applied to the oscilloscope vertical channel. The sweep generator must be capable of sweeping over the entire RF range. If maximum accuracy is desired, a marker generator must also be used.

1. Connect the equipment as shown in Fig. 11-11.

2. Place the oscilloscope in operation (Chapter 5). Switch off internal recurrent sweep. Set the oscilloscope sweep selector and sync selector to external.

3. Place the sweep generator in operation as described in its instruction manual. Switch sweep generator blanking control on or off as desired. Adjust the sweep generator to cover the complete RF range. The center frequency depends upon the receiver. Usually, an

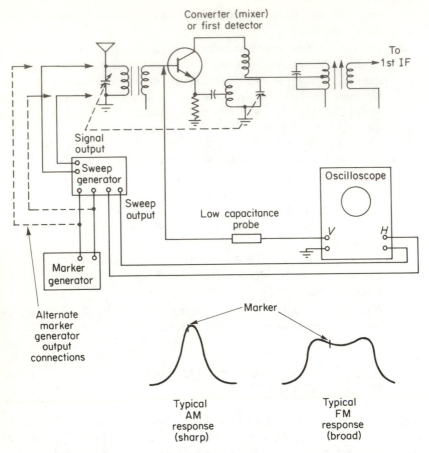

FIGURE 11-11 Alignment of RF (front-end) stages with an oscilloscope.

AM receiver requires a 30-kHz sweep width, and an FM receiver needs about 300 kHz.

4. Check the RF response curve appearing on the oscilloscope against those of Fig. 11–11, or against the receiver specifications.

5. If it is desired to determine the exact frequency at which RF response occurs, the marker generator can be adjusted until the marker pip is aligned at the point of interest. The frequency, or band of frequencies, can be read from the marker generator frequency dial.

6. The amplitude of any point of the response curve can be measured directly on the oscilloscope (assuming that the vertical system is voltage-calibrated).

7. Adjust the RF alignment controls to produce the desired response curve, as specified in the receiver service data. Usually, the RF response of an AM receiver is similar to that of Fig. 11–11b, whereas an FM receiver has a broad response similar to that of Fig. 11–11c.

11–12. ALIGNMENT OF FM DETECTOR WITH AN OSCILLOSCOPE

The detector of an FM receiver (either discriminator or ratio detector) can be aligned using the sweep generator/oscilloscope combination. The test connections are similar to front end and IF alignment. The sweep generator output is connected to the last IF stage input, whereas the oscilloscope vertical channel is connected across the FM detector load resistor. The sweep generator must be capable of sweeping over the entire IF range. If maximum accuracy is desired, a marker generator must also be used.

1. Connect the equipment as shown in Fig. 11–12.
2. Place the oscilloscope in operation (Chapter 5). Switch off internal recurrent sweep. Set the oscilloscope sweep selector and sync selector to external.
3. Place the sweep generator in operation as described in its instruction manual. Switch sweep generator blanking control on or off as desired. Set the sweep generator frequency to the receiver intermediate frequency (usually 10.7 MHz). Adjust the sweep width to about 300 kHz.
4. Check the detector response curve appearing on the oscilloscope against that of Fig. 11–12b, or against the receiver specifications.
5. Adjust the last IF stage and detector alignment controls so that peaks 2 and 4 of the response curve are equal in amplitude above and below the zero line. Also, points 1, 3, and 5 of the response curve should be on the zero reference line.
6. If it is desired to determine the exact frequency at which detector response occurs, the marker generator can be adjusted until the marker pip is aligned at the point of interest. The frequency, or band of frequencies, can be read from the marker generator frequency dial.
7. The amplitude of any point on the response curve can be measured directly on the oscilloscope (assuming that the vertical system is voltage-calibrated).

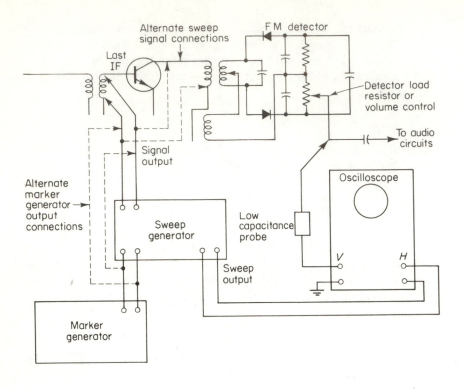

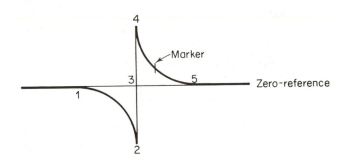

FIGURE 11-12 Alignment of FM detector with an oscilloscope.

11-13. CHECKING RECEIVER AUDIO CHANNELS WITH AN OSCILLOSCOPE

The audio channel of a receiver is essentially an audio amplifier. As such, the audio channel can be checked for frequency response, power output, noise and hum, distortion, and phase shift, as described in Chapter 10. An oscilloscope can also be used as an audio signal tracer for the audio channel.

11-14. SIGNAL-TRACING RECEIVER CIRCUITS WITH AN OSCILLOSCOPE

Since the sweep generator/oscilloscope combination provides complete alignment of the RF, IF, and detector stages, the same combination can be used as signal tracer for these stages. An oscilloscope can be used with an RF signal generator (nonsweep) as a signal tracer. In this case, the oscilloscope is acting essentially as a voltmeter that also displays the signal waveform. An oscilloscope can be used to signal trace any stage, provided that the vertical channel of the oscilloscope will pass the frequency range. If the oscilloscope frequency range is too narrow, a demodulator or RF probe can be used. In that event, the reference signal applied to the receiver must be modulated.

12

Checking Industrial Devices

The oscilloscope has become the "eyes and ears" of the industrial laboratory technician. Confronted with the problem of checking physical quantities or performance testing, a sophisticated technician finds no substitute for a high-quality oscilloscope. This chapter deals with the tests and procedures where a precision laboratory oscilloscope is usually required. Many of the tests can be performed with relatively simple oscilloscopes. The tests and procedures of this chapter have been selected after repeated screening of material. Resourceful laboratory technicians can adapt many of the basic procedures to other uses.

12-1. GENERAL PULSE AND SQUARE-WAVE MEASUREMENT TECHNIQUES

The great majority of industrial oscilloscope applications involve the measurement of pulses and square waves. For that reason, the following sections review pulse measurement techniques, as well as pulse characteristics.

12-1-1.   Pulse Definitions

The following terms are commonly used in describing pulse characteristics. The same terms are used with square waves. The terms are illustrated as applied in Fig. 12-1. The input pulse represents an ideal input waveform for comparison. The other waveforms in Fig. 12-1 represent typical output waveforms in order to show the relationships. The terms are defined as follows:

Rise time, T_R: the time interval during which the amplitude of the output voltage changes from 10% to 90% of the rising portion of the pulse.

Fall time, T_F: the time interval during which the amplitude of the output voltage changes from 90% to 10% of the falling portion of the waveform.

Pulse width (or duration), T_W: the time duration of the pulse measured

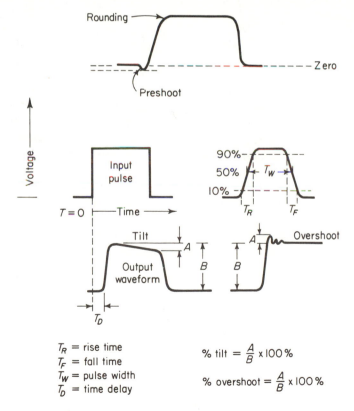

T_R = rise time

T_F = fall time

T_W = pulse width

T_D = time delay

% tilt $= \dfrac{A}{B} \times 100\%$

% overshoot $= \dfrac{A}{B} \times 100\%$

FIGURE 12-1   Basic pulse and square-wave definitions.

between the 50% amplitude levels of the rising and falling portions of the waveform.

Time delay, T_D: the time interval between the beginning of the input pulse (T = 0), and the time when the rising portions of the output pulse attains an arbitrary amplitude of 10% above the baseline.

Tilt: a measure of the tilt of the full amplitude; flat-top portion of a pulse. The tilt measurement is usually expressed as a percentage of the amplitude of the rising portion of the pulse.

Overshoot: a measure of the overshoot occurring generally above the 100% amplitude level. This measurement is also expressed as a percentage of the pulse rise.

These definitions are only guides. When the actual pulses are very irregular (such as excessive tilt, overshoot, etc.), the definitions may become ambiguous. In such cases, a more complete description of the pulse will probably be necessary.

12-1-2. Rule of Thumb for Rise-Time Measurements

Since rise-time measurements are of special importance in pulse testing, the relationship between the oscilloscope rise time, and the rise time of the device under test must be taken into account. Obviously, the accuracy of rise-time measurements can be no greater than the rise time of the oscilloscope. Also, if the device is tested by means of an external pulse from a pulse generator, the rise time of the pulse generator must also be taken into account.

For example, if an oscilloscope with a 20-ns rise time is used to measure the rise time of a 15-ns device, the measurements would be hopelessly inaccurate. If a 20-ns pulse generator and a 15-ns oscilloscope were used to measure the rise time of a device, the fastest rise time for accurate measurement would be something greater than 20 ns. Two basic rules of thumb can be applied to rise-time measurements.

The first method is known as the *root of the sum of the squares*. It involves finding the squares of all the rise times associated with the test, adding these squares together, and then finding the square root of this sum. For example, using the 20-ns pulse generator and the 15-ns oscilloscope, the calculation would be as follows:

$$20 \times 20 = 400 \qquad 15 \times 15 = 225 \qquad 400 + 225 = 625$$

$$\sqrt{625} = 25 \text{ (ns)}$$

This means that the fastest possible rise time capable of measurement is 25 ns.

One major drawback to this rule is that the coaxial cables required to interconnect the test equipment are subject to "skin effect." As frequency increases, the signals tend to travel on the outside or skin of the conductor. This decreases conductor area, and increases resistance. In turn, this increases cable loss. The losses of cables do not add properly to apply the root-sum-squares method, except as an approximation.

The second rule or method states that if the equipment or signal being measured has a rise time *10 times* slower than the test equipment, the error is 1%. This amount is small and can be considered as negligible. If the equipment being measured has a rise time *3 times* slower than the test equipment, the error is slightly less than 6%. By keeping these relationships in mind, the results can be interpreted intelligently.

12-1-3. Matching Pulse Generator Impedance to Device under Test

One problem often encountered when testing pulsed equipment is the matching of impedances. To provide a smooth transition between devices of different characteristic impedance, each device must encounter a total impedance equal to its own characteristic impedance. A certain amount of signal attenuation is usually required to achieve this transition. A simple resistive impedance-matching network that provides minimum attenuation is shown in Fig. 12-2, together with the applicable equations.

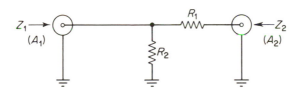

To match impedances: $R_1 = \sqrt{Z_2(Z_2 - Z_1)}$ $R_2 = Z_1\sqrt{\dfrac{Z_2}{Z_2 - Z_1}}$

Voltage attenuation seen from Z_1 end (A_1): $A_1 = \dfrac{R_1}{Z_2} + 1$

Voltage attenuation seen from Z_2 end (A_2): $A_2 = \dfrac{R_1}{R_2} + \dfrac{R_1}{Z_1} + 1$

FIGURE 12-2 Resistive impedance matching network for pulse circuits.

For example, to match a 50-Ω system to a 125-Ω system,

$$Z_1 = 50 \ \Omega \quad \text{and} \quad Z_2 = 125 \ \Omega$$

Therefore,

$$R_1 = \sqrt{125(125 - 50)} = 96.8 \ \Omega$$

$$R_2 = 50 \sqrt{\frac{125}{125 - 50}} = 64.6 \ \Omega$$

Though the network in Fig. 12–2 provides minimum attenuation for a purely resistive impedance-matching device, the attenuation as seen from one end does not equal that seen from the other end. A signal applied from the lower-impedance source (Z_1) encounters a voltage attenuation (A_1) that may be determined as follows:

Assume that R_1 is 96.8 Ω and Z_2 is 125 Ω.

$$A_1 = \frac{96.8}{125} + 1 = 1.77 \text{ attenuation}$$

A signal applied from the higher-impedance source (Z_2) will produce an even greater voltage attenuation (A_2) that may be determined as follows:

Assume that $R_1 = 96.8 \ \Omega$, $R_2 = 64.6 \ \Omega$, and impedance $Z_1 = 50 \ \Omega$.

$$A_2 = \frac{96.8}{64.6} + \frac{96.8}{50} + 1 = 4.44 \text{ attenuation}$$

12–2. MEASUREMENT OF PULSE TIME WITH EXTERNAL TIMING PULSES

On those oscilloscopes where the horizontal sweep circuits are calibrated in units of time, pulse width (duration) or spacing between pulses can be measured directly by counting the number of screen divisions along the horizontal axis. This method is described in Chapter 7, as is the method for converting frequency-calibrated horizontal sweeps to time-calibrated sweeps (Secs. 7–1 through 7–3).

It is also possible to use an external *time-mark generator* to calibrate the horizontal scale of an oscilloscope for time measurement. A time-mark generator produces a pulse-type timing wave which is a series of sharp spikes, spaced at precise time intervals. These pulses are applied to the vertical input and appear as a wavetrain as shown in Fig. 12–3. The oscilloscope horizontal gain and positioning controls are adjusted to align the timing spikes with

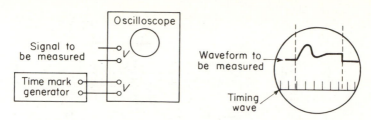

FIGURE 12-3 Typical connections and pattern for time measurement with external timing pulses.

screen lines, until the screen divisions equal the timing pulses. The accuracy of the oscilloscope timing circuits is then of no concern, since the horizontal channel is calibrated against the external time-mark generator. The timing pulses can be removed, and the signal to be measured applied to the vertical input, provided that the horizontal gain and positioning controls are not touched. Duration or time is read from the calibrated screen divisions in the normal manner. If the oscilloscope has a dual-trace feature, the time-mark generator can be connected to one vertical input; the other vertical input receives the signal to be measured. The two traces can be superimposed, or aligned, whichever is convenient.

Time-mark generators are used for many reasons, but their main advantages are greater accuracy and resolution. For example, a typical time-mark generator unit will produce timing signals at intervals of 10, 1, and 0.1 μs.

12-3. MEASUREMENT OF PULSE DELAY

The time interval (or delay) between an input pulse and output pulse introduced by a delay line, digital circuit, multivibrator, or similar circuit, can be measured on an oscilloscope with dual trace. If the delay is exceptionally short, the screen divisions can be calibrated with an external time-mark generator. If the oscilloscope has three vertical inputs, the timing wave from the time-mark generator can be displayed simultaneously with the input and output pulses.

1. Connect the equipment as shown in Fig. 12-4.
2. Place the oscilloscope in operation (Chapter 5). Switch on the internal recurrent sweep. Set the sweep selector and sync selector to internal.
3. Switch on the delay device, if it is powered.
4. Switch on the time-mark generator and pulse generator as described in their instruction manuals.

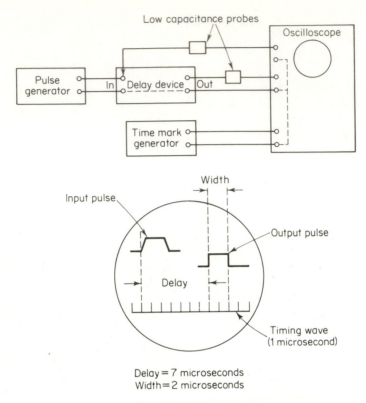

Delay = 7 microseconds
Width = 2 microseconds

FIGURE 12-4   Measurement of pulse delay.

5. If the oscilloscope is a multiple-trace instrument, set the sweep frequency and sync controls for a single, stationary input pulse and output pulse as shown in Fig. 12-4.

6. Set the horizontal and vertical gain controls and pulse generator output for desired pulse pattern width and height.

7. Count the timing spikes between the input and output pulses to determine the delay interval. Count the timing spikes between the beginning and end of each pulse to determine pulse width or duration.

NOTE

If the oscilloscope is a dual-trace instrument, the screen divisions must be calibrated against the time-mark generator as described in Sec. 12-2. The time-mark generator can then be removed and the input and output pulses applied to the two vertical channels.

12-4. MEASURING IMPEDANCE WITH A PULSE GENERATOR AND OSCILLOSCOPE

A pulse generator/oscilloscope combination can be used to measure impedance of an unknown device by comparison of the reflected pulse with the incident pulse.

1. Connect the equipment as shown in Fig. 12-5.

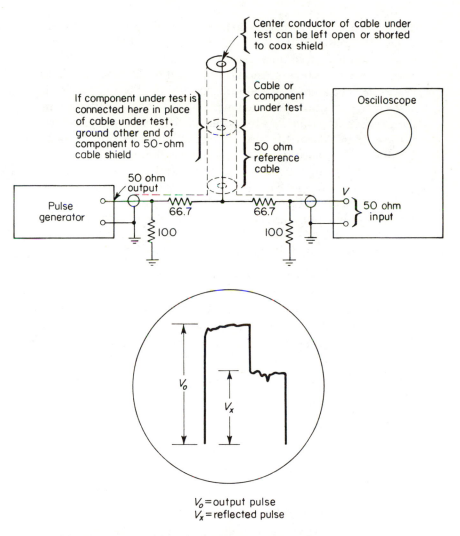

V_o = output pulse
V_x = reflected pulse

FIGURE 12-5 Measuring impedance with pulse generator and oscilloscope (using output and reflected pulses).

2. Place the oscilloscope in operation (Chapter 5). Switch on the internal recurrent sweep. Set the sweep selector and sync selector to internal.

3. Switch on the pulse generator as described in its instruction manual. Set the sweep frequency and sync controls to display the output pulse, and the first reflected pulse.

NOTE

This impedance measurement method is based on the comparison of reflected pulses with output pulses. As a signal travels down a transmission line, each time it encounters a mismatch or different impedance, a reflection is generated and sent back along the line to the source. The amplitude and polarity of the reflection are determined by the value of the impedance encountered in relation to the characteristic impedance of the cable. If the mismatch impedance is higher than that of the line, the reflection will be of the same polarity as the applied signal; if it is lower than that of the line, the reflection will be of opposite polarity. The reflected signal is added to or subtracted from the amplitude of the pulse if it returns to the source before the pulse has ended. Thus, for a cable with an open end (no termination), the impedance is infinite and the pulse amplitude would be doubled. For a cable with a shorted end, the impedance is zero and the pulse would be canceled.

4. Observe the output and reflected pulses on the oscilloscope screen. Using Fig. 12–5 as a guide, determine the values of V_O (output voltage amplitude) and V_X (reflected voltage amplitude).

5. Calculate the unknown impedance using the equation

$$Z = \frac{50}{(2V_O/V_X) - 1}$$

where Z = unknown impedance

      50 = reference impedance (50-Ω coax line)

      V_0 = peak amplitude produced by the 50-Ω reference impedance

      V_X = peak amplitude at the time of the reflection

12-5. MEASURING STRAIN WITH AN OSCILLOSCOPE

An oscilloscope provides a reliable means of measuring dynamic strain (where strain changes rapidly with time). Static strain (where the strain remains constant) can be measured easily with strain gauges connected in a bridge circuit. When the strain varies over a short time span, the simple bridge circuit cannot record this action, as an oscilloscope can. Because of the instantaneous nature of the trace, the display should be photographed (unless a storage-type oscilloscope is used).

As shown in Fig. 12-6, the strain gauges are connected in a bridge circuit, the output of which is applied to the oscilloscope vertical channel. The bridge is balanced under no-strain conditions with potentiometer R_1. One strain gauge is placed on the material or structure to be tested; the other identical strain gauge is used as a reference. The oscilloscope is not deflected when the bridge is balanced (no-strain). When the material or structure is stressed,

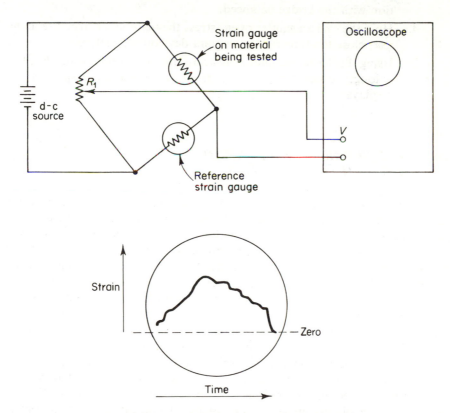

FIGURE 12-6 Measuring strain with oscilloscope.

the resistance of the attached strain gauge is changed. This unbalances the bridge and produces a d-c output that is proportional to the change. The d-c output deflects the oscilloscope vertical trace, and produces a plot of strain versus time. The bridge circuit can be calibrated in terms of strain (micro-inches of variation per inch-ounces of applied force), or strain versus voltage deflection, or whatever proves convenient for the particular test. The entire trace can be photographed for a permanent record.

1. Connect the equipment as shown in Fig. 12–6.

2. Place the oscilloscope in operation (Chapter 5). Switch on the internal recurrent sweep. Set the sweep selector and sync selector to internal. Use a sweep time interval that will be longer than the strain interval. Set up the oscilloscope camera as necessary.

3. Set the oscilloscope to measure dc. Balance the bridge adjusting potentiometer R_1. The oscilloscope should be at zero vertical deflection with the bridge balanced.

4. Hold the camera shutter open, stress the material or structure under test, close the camera shutter, and develop the picture.

5. Using the developed photo, measure the strain versus time plot. Use a longer sweep time interval if the complete strain plot is not displayed.

12–6. MEASURING ACCELERATION WITH AN OSCILLOSCOPE

An oscilloscope will provide a means of measuring variations in acceleration as a function of time. The procedure is almost identical to that of dynamic strain measurement (Sec. 12–5). The major difference is that a resistance-type accelerometer is used in place of the strain gauge (Fig. 12–7). The opposite leg of the bridge is composed of a fixed resistance to match the accelerometer resistance. The bridge is balanced under no-acceleration conditions, or at some preselected value of acceleration by potentiometer R_1. The oscilloscope is not deflected when the bridge is balanced (no acceleration). When the acceleration changes, the accelerometer resistance is changed. This unbalances the bridge and produces a d-c output that is proportional to acceleration change. The d-c output deflects the oscilloscope vertical trace and produces a plot of acceleration change versus time. The bridge can be calibrated in terms of acceleration (feet per second^2 per volt) or whatever is convenient. The entire trace can be photographed for a permanent record. (Set up the equipment as shown in Fig. 12–7, and follow the procedure of Sec. 12–5.)

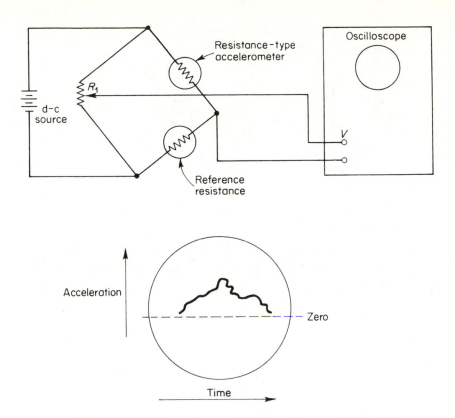

FIGURE 12-7 Measuring acceleration with an oscilloscope.

12-7. MEASURING PRESSURE
WITH AN OSCILLOSCOPE

An oscilloscope will provide a means of measuring variations in pressure as a function of time. The procedure is almost identical to that of dynamic strain measurement (Sec. 12-5). The major difference is that a resistance-type pressure gauge is used in place of the strain gauge (Fig. 12-8a). These pressure gauges are actuated by a bellows or diaphragm that moves with pressure changes. Bellows or diaphragm movement causes a corresponding change in resistance. The opposite leg of the bridge is composed of a fixed resistance to match the pressure gauge resistance. The bridge is balanced under no-pressure conditions, or at some preselected value of pressure, by potentiometer R_1. The oscilloscope is not deflected when the bridge is balanced (no pressure). When pressure changes, the pressure gauge resistance changes. This unbalances the bridge and produces a d-c output that is proportional to pressure change. The

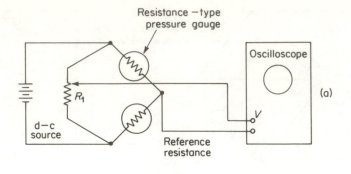

Resistance –type
pressure gauge

Oscilloscope

(a)

R_1

d–c
source

Reference
resistance

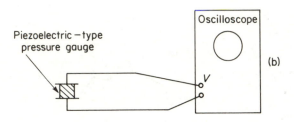

Piezoelectric –type
pressure gauge

Oscilloscope

(b)

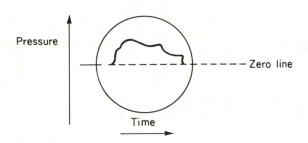

Pressure

Zero line

Time

FIGURE 12-8 Measuring pressure with an oscilloscope.

d-c output deflects the oscilloscope vertical trace and produces a plot of pressure change versus time. The bridge can be calibrated in terms of pressure (pounds per volt) or whatever is convenient.

Pressure is sometimes measured with piezoelectric pressure pickups, as shown in Fig. 12–8b. These pickups are self-generating and do not require a bridge circuit. Piezoelectric pickups are connected directly to the vertical input of the oscilloscope.

Set up the equipment as shown in Fig. 12–8a or b, and follow the procedure of Sec. 12–5.

12-8. MEASURING ELECTRICAL NOISE WITH AN OSCILLOSCOPE

An oscilloscope provides a reliable means of measuring electrical noise or "hash." Such electrical noise can be defined as any undesired, nonrepetitive signal present in a circuit, and includes background noise, transient noise, etc. Usually, electrical noise is a combination of many frequencies and waveforms, all of which is usually of major importance in any test. This can be measured on an oscilloscope in the normal manner. Because of the instantaneous nature of the trace, the display should be photographed (unless a storage-type oscilloscope is used; refer to Sec. 9–11).

As shown in Fig. 12–9, the noise source is connected to the oscilloscope vertical channel. The source can be connected directly to the vertical input, or through a tuned amplifier. The tuned amplifier (such as is found in sound and

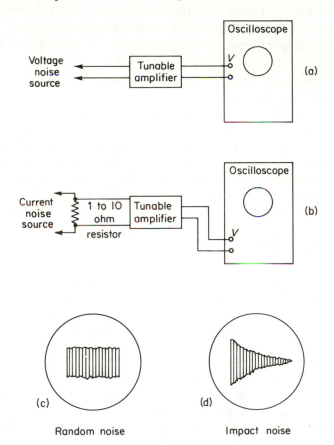

FIGURE 12-9 Measuring electrical noise with an oscilloscope.

vibration analyzers, wave analyzers, etc.) is used where noise frequency is of particular importance. If the noise source is a voltage, the vertical input can be connected across a component or branch of the circuit. If the noise appears as a current, a resistance must be inserted in the circuit, and the voltage drop across the resistance measured. If it is not practical to interrupt the circuit, a current probe can be used (Sec. 6–10).

1. Connect the equipment as shown in Fig. 12–9.

NOTE

The tuned amplifier can be omitted if peak noise amplitude is the only factor of interest.

2. Place the oscilloscope in operation (Chapter 5). Switch on the internal recurrent sweep. Set the sweep selector and sync selector to internal. Set up the oscilloscope camera as necessary.
3. Set the sweep frequency and horizontal and vertical gain controls to display the noise pattern similar to that of Fig. 12–9c or d.
4. Measure the peak-to-peak noise amplitude along the voltage-calibrated vertical axis.
5. If the noise pattern appears repetitive, measure the time interval between noise intervals, using the time-calibrated horizontal axis.
6. If impact noise is to be measured and recorded, hold the camera shutter open, initiate the impact, close the camera shutter, and develop the picture. Using the developed photo, measure the peak-to-peak amplitude, as well as the time interval of the impact noise.
7. If it is desired to determine the frequency range of noise signals, tune the amplifier to each frequency of interest and note the time and amplitude of the noise signals.

12–9.   *MEASURING ACOUSTIC NOISE AND SOUND WITH AN OSCILLOSCOPE*

The procedure for measuring acoustic noise (or sound) with an oscilloscope is almost identical to that for electrical noise (Sec. 12–8). The major difference is that a microphone is used as the noise pickup, instead of connecting into a circuit to measure noise voltage or current. The microphone acts as a transducer and converts acoustic noise into an electrical signal. Usually a capacitor-type microphone is used, with a preamplifier, for most acoustic noise measurements. Also, the voltage readings on the oscilloscope vertical channel can be converted to decibels, if desired.

Set up the equipment as shown in Fig. 12–10, and follow the procedure of Sec. 12–8.

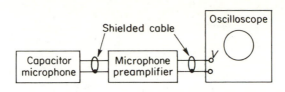

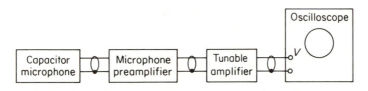

FIGURE 12-10 Measuring acoustic noice (or sound) with an oscilloscope.

12-10. MEASURING VIBRATION WITH AN OSCILLOSCOPE

The procedure for measuring vibration (continuous, random, or impact) with an oscilloscope is almost identical to that for electrical noise (Sec. 12–8) or acoustic noise (Sec. 12–9). The major difference is that a vibration pickup is used instead of a microphone. The pickup acts as a transducer and converts vibration into an electrical signal. Most vibration transducers are piezoelectric instruments which produce an a-c voltage proportional to acceleration of the vibrating body. Some vibration transducers contain integration networks which provide output voltages proportional to velocity and displacement. Other vibration transducers are supplied with special preamplifiers. The vibration transducer manufacturers often supply calibration data which permit the voltage readings on the oscilloscope to be converted to a direct readout of velocity, acceleration, displacement, amplitude, etc.

(Set up the equipment as shown in Fig. 12–11 and follow the procedure of Sec. 12–8.)

12-11. MEASUREMENT OF ROTATIONAL SPEED WITH AN OSCILLOSCOPE

An oscilloscope can be used to measure rotational speed of machinery. In this application, the oscilloscope functions as an indicating device. The actual rotation is converted to an electrical signal by a transducer. There are three basic types of transducers: magnetic, capacitive, and photoelectric. The magnetic tachometer-type transducer is the most common. Such magnetic

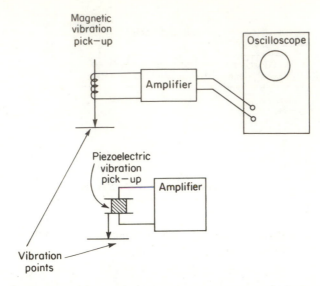

FIGURE 12-11 Measuring vibration with an oscilloscope.

transducers are miniature generators driven by the rotating machinery. Usually they produce an a-c output. Some magnetic transducers produce an output voltage proportional to rotational speed and are rated in rpm per volt. Other magnetic transducers produce a signal, the frequency of which is proportional to rotational speed. The capacitive and photoelectric transducers are almost always of the frequency type.

12-11-1. Procedure for Frequency-type Speed Transducers

1. Connect the equipment as shown in Fig. 12–12.
2. Place the oscilloscope in operation (Chapter 5). Switch off the internal recurrent sweep. Set the sweep selector and sync selector to external.
3. Connect the transducer to the machinery as described in the transducer instruction manual. Usually, capacitive and photoelectric transducers do not require direct coupling.
4. Place the audio generator in operation as described in its instruction manual.
5. Adjust the audio generator frequency to obtain a stable pattern on the oscilloscope. Identify the transducer output frequency by means of Lissajous figures, as described in Sec. 7–4.
6. Convert the frequency measurement into a speed indication, using the conversion factor supplied with the transducer.

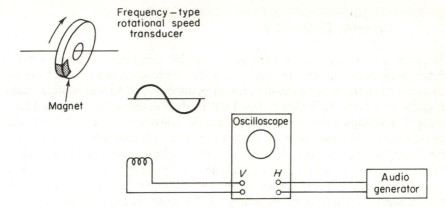

FIGURE 12-12 Measuring rotational speed with frequency-type speed transducer.

7. If desired, any of the frequency measurement techniques described in Chapter 7 can be used instead of Lissajous figures.

12-11-2. Procedure for Voltage-type Speed Transducers

1. Connect the equipment as shown in Fig. 12–13.

2. Place the oscilloscope in operation (Chapter 5). Switch on the internal recurrent sweep. Set the sweep selector and sync selector to internal.

3. Connect the transducer to the machinery as described in the transducer instruction manual.

4. Set the oscilloscope to measure ac or dc, depending upon the transducer output.

5. With the machinery operating, measure the transducer voltage output on the oscilloscope vertical scale.

6. Convert the voltage measurement into a speed indication, using the conversion factor supplied with the transducer.

FIGURE 12-13 Measuring rotational speed with voltage-type speed transducer.

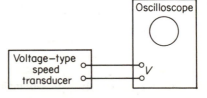

12-12. MEASUREMENT OF TRANSMISSION-LINE CHARACTERISTICS

An oscilloscope and pulse generator can be combined to measure the characteristics of transmission lines, including coaxial cable lines. Such an arrangement is known as a time-domain reflectometer (TDR) and was originally developed by Hewlett-Packard. The TDR is often described as a "closed-loop radar." A voltage pulse or step is propagated down a transmission line under investigation, and the incident (outgoing) and reflected voltage waves are monitored by the oscilloscope at a particular point on the line. This "echo" technique reveals at a glance the characteristic impedance of the line, as well as both the position and nature (resistive, inductive, or capacitive) of any discontinuity on the line. Time-domain reflectometers also indicate whether losses in a transmission system are series losses or shunt losses.

12-12-1. Basic Time-Domain Reflectometer Operation

The basic time-domain reflectometer circuit is shown in Fig. 12–14. In operation, the pulse generator produces a positive-going wave or "step" that is fed into the transmission line under test. The step wave travels down the transmission line under test, at a velocity determined by the type of transmission line. If the load impedance is exactly equal to the characteristic im-

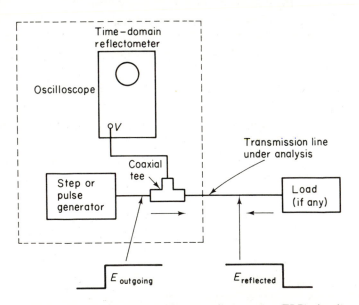

FIGURE 12-14 Basic time-domain reflectometer (TDR) circuit.

pedance of the line, no wave is reflected, and all that is seen on the oscillo-scope is the outgoing pulse recorded as the wave passes the point on the line monitored by the oscilloscope (Fig. 12–15a). If a mismatch exists at the load (the load impedance does not match the line impedance exactly), part of the outgoing wave is reflected. The reflected pulse voltage appears on the oscilloscope, algebraically added to the outgoing wave (Fig. 12–15b).

12-12-2. Analyzing Time-Domain Reflectometer Displays

The shape of a waveform on a time-domain reflectometer requires analysis. The shape and amplitude of the reflected wave in relation to the outgoing wave reveal the nature and magnitude of mismatches, loads, discontinuities, and so on, on the line.

Figure 12–16 summarizes typical TDR displays and the corresponding load or mismatch responsible for the displays. Note that the displays shown in Fig. 12–16 are *theoretical*, and that actual displays will never be quite as precise.

In Fig. 12–16a (an open-circuit termination where the load impedance is infinity, such as an open-ended coaxial cable), the full voltage is reflected back and added to the outgoing voltage.

In Fig. 12–16b (a short-circuit termination where the load impedance is zero, such as a coaxial cable with the inner and outer conductors shorted at the free end), the voltage drops to zero and cancels the outgoing voltage.

In Fig. 12–16c (a pure resistive load where the load impedance is twice the impedance of the transmission line, such as a 50-Ω coaxial cable with a 100-Ω terminating resistor), one-third of the voltage is reflected back and added to the outgoing voltage.

In Fig. 12–16d (a pure resistive load where the load impedance is one-half the impedance of the transmission line, such as a 50-Ω coaxial cable with

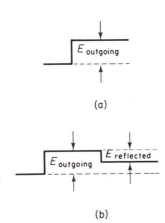

FIGURE 12-15 Basic TDR patterns. (a) Reflected signal zero; (b) reflected signal added algebraically to outgoing signal (mismatch).

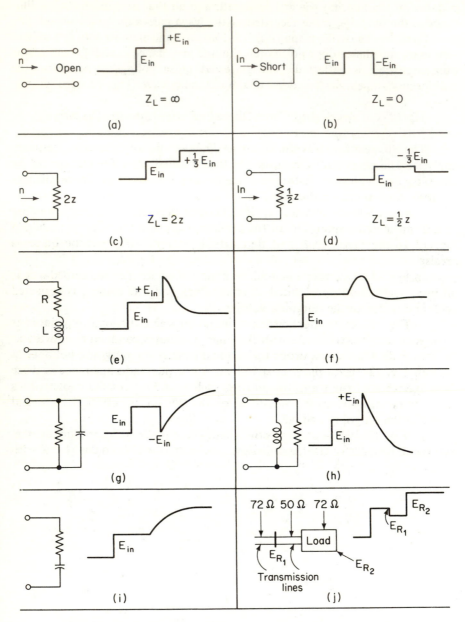

FIGURE 12–16 Typical TDR displays and corresponding condition responsible for displays.

a 25-Ω terminating resistor), one-third of the voltage is reflected back and cancels one-third of the outgoing voltage.

Note that the reflected pulses from an open, a short, or a pure resistive load are the same shape as those of the pulse generator, and they are added algebraically to the outgoing pulse on the oscilloscope display. Unlike pure resistive loads, capacitive or inductive loads change the shape of the reflected pulse.

In Fig. 12–16e (a load combination of resistance and inductance in series), the leading edge of the reflected wave is the same shape as the outgoing pulse and is added algebraically. With time, the pulse slopes off to a value below that of the outgoing pulse. This slope is caused by the effect of the inductance. When the pulse waveform is first applied, the inductance opposes current flow and appears as an infinite impedance. As current starts to flow, the impedance begins to drop, and the voltage slopes off in proportion. Current flow is limited only by the effect of the resistance.

The display of Fig. 12–16e appears only for very large values of inductance and resistance. If the values are small (particularly the inductance value), the display is more like that of Fig. 12–16f. This is because the time constant of the reflected wave is so short that the wave decays to the final value almost before the TDR oscilloscope can rise. This problem is minimized by the use of a sampling oscilloscope (Sec. 1–7).

In Fig. 12–16g (a load combination of capacitance and resistance in parallel or shunt), the capacitor acts like a short to the pulse waveform. Initially, the load impedance is zero, the voltage drops to zero and cancels the outgoing voltage. With time, the voltage builds up across the capacitor, and current flow is reduced. The rate of capacitor charge and discharge is determined by the RC circuit values.

In Fig. 12–16h (a load combination of resistance and inductance in parallel or shunt), the leading edge of the reflected wave is the same shape as the outgoing pulse and is added algebraically. With time, the pulse slopes off to zero. This is a similar reaction to that of a resistance–inductance in series. However, since the inductance is in shunt, the current flow is not limited by the resistance. Therefore, the voltage drops to zero.

In Fig. 12–16i (a load combination of capacitance and resistance in series), the capacitor initially appears as a short to the pulse leading edge, leaving only the resistance. With time, the voltage builds up across the capacitor, and is added algebraically. The rate of capacitor charge and discharge is determined by the RC circuit values.

In Fig. 12–16j (multiple mismatches, or mismatches and a load), the reflections must be analyzed separately. The mismatch at the junction of the two transmission lines (72 to 50 Ω) generates a reflected wave E_{R1}. Similarly, the mismatch at the load (50 to 72 Ω) creates a reflection E_{R2}.

By studying Fig. 12–16, it is seen that the reflected waveshape determines the type of mismatch or load present on the transmission line being tested. Once these waveshapes are learned, both the type and location of the mismatch can be determined. If the amplitude and rate of slope could be measured accurately, the actual values of resistance, inductance, and capacitance causing the particular waveshape could be calculated. However, such analysis and calculations require a study of complex waveforms and a beyond the scope of this publication.

12-12-3. Locating Mismatches or Discontinuities on Lines

In addition to determining the characteristics of mismatches or loads in transmission lines, a TDR can be used to locate mismatches (physically) on lines.

The basic procedure for locating mismatches is to measure the travel time and speed of the step or pulse as it passes down the transmission line and then back to the oscilloscope. This is done by measuring the distance (in time) between the outgoing and reflected waves (using the oscilloscope horizontal scale in a manner similar to that for measuring pulse delay as described in Sec. 12-3). The reflected wave is readily identified since it is separated in time from the outgoing wave. Make certain to measure time between the *same points* on both waves. Generally, the leading edge of both waves is a convenient point for measurement of time.

By measuring both the time and velocity of propagation, the distance can be calculated by

$$\text{distance} = \text{velocity of propagation} \times \frac{T}{2}$$

where T is the transit time from monitoring point to the mismatch and back again, as measured on the oscilloscope horizontal scale.

If the velocity of propagation for a particular type of transmission line is not known, it can be found from an experiment on a known length of the same type of line. For example, the time required for the outgoing wave to travel down, and the reflected wave to travel back from an open-circuit termination at the end of a 120-cm length of RG-9A coaxial cable is 11.4 ns. This means that the velocity of propagation is 21 cm/ns ($11.4/2 = 5.7$; $120/5.7 = 21$).

12-13. MEASURING POWER SUPPLY RIPPLE WITH AN OSCILLOSCOPE

An oscilloscope provides the only satisfactory means of measuring ripple in power supplies. An oscilloscope will display the ripple waveform from which the amplitude, frequency, and nature of the ripple voltage can be determined.

Usually, the oscilloscope is set to the a-c mode when measuring ripple, since this blocks the d-c output of the power supply. Normally, ripple voltage is small in relation to the power supply voltage. Therefore, if the oscilloscope gain is set to display the ripple, the power supply d-c voltage would drive the display off screen.

1. Connect the equipment as shown in Fig. 12–17.
2. Place the oscilloscope in operation (Chapter 5). Switch on the internal recurrent sweep. Set the sweep selector and sync selector to internal.
3. Switch on the power supply. If the test is to be made under load conditions, close switch S_1. Open switch S_1 for a no-load test. The value of R_1 must be chosen to provide the power supply with the desired load. Usually, power supply ripple is tested at full load and half load.

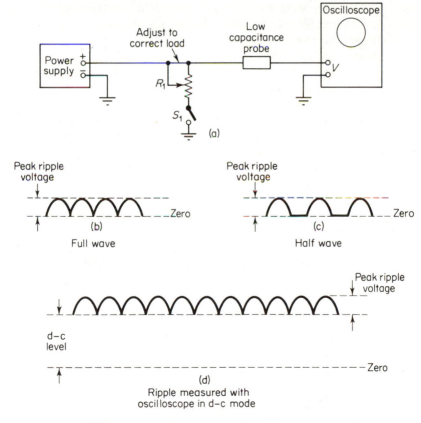

FIGURE 12-17 Measuring power supply ripple with an oscilloscope.

NOTE

Make certain that the combined d-c and ripple output of the power supply does not exceed the oscilloscope voltage input rating.

4. Adjust the sweep frequency and sync controls to produce two or three stationary cycles of each wave on the screen.
5. Measure the peak amplitude of the ripple on the voltage-calibrated method described in Chapter 7.

NOTE

When measuring frequency, note that a full-wave power supply (Fig. 12–17b) will produce two ripple "humps" per cycle, whereas a half-wave power supply (Fig. 12–17c) will produce one "hump" per cycle.

6. If the power supply d-c output must be measured simultaneously with the ripple, set the oscilloscope for d-c mode. The baseline should be deflected upward to the d-c level, and the ripple should be displayed at that level (Fig. 12–17d). If this drives the display off screen, reduce the vertical gain using the step attenuator, and measure the d-c output. Then return the vertical gain to a level where the ripple can be measured and use the vertical position control to bring the display back onto the screen for ripple measurement. This procedure will work with most laboratory oscilloscopes. On shop-type oscilloscopes where the vertical gain control must be set at a given "calibrate" position, the procedure will prove difficult, if not impossible.
7. A study of the ripply waveform can reveal defects in the power supply. For example, if the power supply is unbalanced (one rectifier passing more current than the other), the ripple "humps" will be unequal in amplitude. If there is noise or fluctuation in the power supply, the ripple "humps" will vary in amplitude or shape. If the ripple varies in frequency, the a-c source is varying. If a full-wave power supply produces a half-wave output, one rectifier is not passing current.

12-14. *MEASURING COMMUTATOR RIPPLE WITH AN OSCILLOSCOPE*

An oscilloscope can be used to measure commutator ripple of a d-c generator. The procedure is similar to that for measurement of a power supply (Sec. 12–13).

1. Connect the equipment as shown in Fig. 12–18.
2. Place the oscilloscope in operation (Chapter 5). Switch on the internal recurrent sweep. Set the sweep selector and sync selector to internal.
3. Energize the generator. If the test is to be made under load conditions, close switch S_1. Open switch S_1 for a no-load test. The value of R_1 must be chosen to provide the generator with the desired load. Usually, generator ripple is tested at full load and half load.

NOTE

Make certain that the combined d-c and ripple output of the generator does not exceed the oscilloscope voltage input rating.

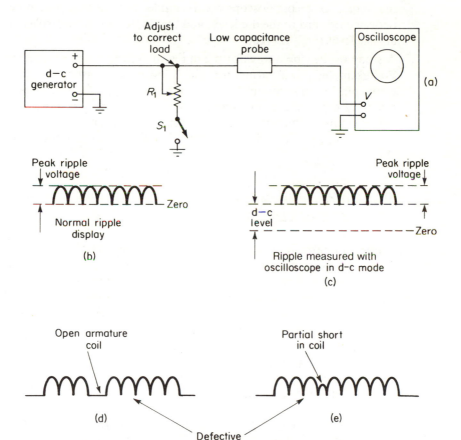

FIGURE 12-18   Measuring generator ripple with an oscilloscope.

4. Adjust the sweep frequency and sync controls to produce two or three stationary cycles of each wave on the screen.

5. Measure the peak amplitude of the ripple on the voltage-calibrated vertical axis (Fig. 12–18b). Measure the ripple frequency using the most convenient method described in Chapter 7.

NOTE

Usually, the ripple frequency of a generator is dependent upon the number of armature turns and the armature speed. For example, if the armature has 10 turns and the armature is driven at 10 revolutions per second (600 rpm), the ripple frequency would be 100 Hz.

6. If the generator d-c output must be measured simultaneously with the ripple, set the oscilloscope for d-c mode. The baseline should be deflected upward to the d-c level, and the ripple should be displayed at that level (Fig. 12–18c).

7. A study of the ripple waveform can reveal defects in the generator. For example, if one armature coil is open or shorted, the corresponding ripple "hump" will be missing (Fig. 12–18d). If one armature coil has a loose connection or high-resistance short, the corresponding "hump" will be lower in amplitude. If there is "brush chatter," the ripple "humps" will vary in amplitude or shape.

12–15. *CHECKING OSCILLATOR CIRCUITS WITH AN OSCILLOSCOPE*

An oscilloscope can provide an excellent means of checking the pulse or square-wave oscillator circuits found in industrial electronic equipment. Such circuits are difficult, if not impossible, to check by any other means. Amplifier and power supply circuits can be checked with a meter, or signal generator and meter combination, but an oscillator or generator output must be monitored as to frequency, amplitude, and waveshape.

It is impossible to describe the procedures for checking each type of oscillator, multivibrator, generator, and so on, in this book. The following procedures are applicable to most self-generating circuits and can be adapted to meet the test needs of specific circuits.

1. Connect the equipment as shown in Fig. 12–19. Use a low-capacitance probe for all measurements. The oscilloscope should have a voltage-calibrated vertical axis, and a time-calibrated horizontal axis.

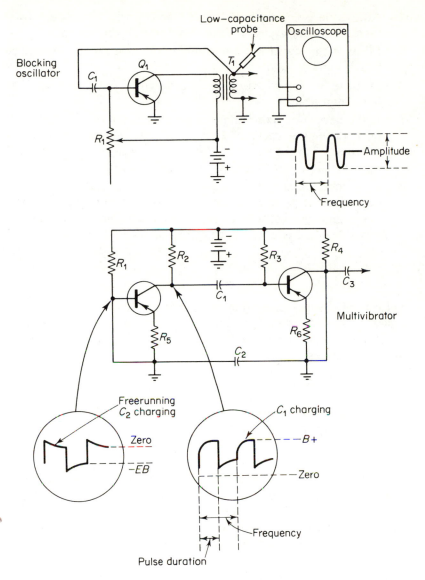

FIGURE 12-19 Checking oscillator circuits with an oscilloscope.

2. Place the oscilloscope in operation (Chapter 5). Switch on the internal recurrent sweep. Set the sweep selector and sync selector to internal.

3. Switch on the oscillator circuit. Touch the low-capacitance probe to the circuit test point. Figure 12–19 shows two typical oscillator circuits and typical waveforms.

4. Adjust the sweep frequency and sync controls to produce two or three stationary cycles of each waveform on the screen.

5. Measure the waveform amplitude of interest on the voltage-calibrated vertical axis.

6. Measure the waveform duration of interest on the time-calibrated horizontal axis.

7. If the horizontal axis is time-calibrated, and it is desired to determine the pulse repetition rate or output frequency, measure the duration of *one complete cycle* (not one pulse), then convert this time duration into frequency as described in Sec. 7–2.

8. If the horizontal axis is frequency calibrated, and it is desired to determine the pulse repetition rate or output frequency, adjust the sweep and sync controls to produce one complete cycle, then convert this into frequency as described in Sec. 7–3.

NOTE

The other frequency measuring techniques of Chapter 7 (Lissajous figures, Z-axis modulation) could be used. A time-calibrated horizontal axis is usually the most convenient for measurement of pulse frequencies.

12–16. USING AN OSCILLOSCOPE IN DIGITAL EQUIPMENT TESTING

Because of the high speeds involved, and the multiple-signal nature of digital devices, the *logic analyzer* (or *logic signal analyzer*) is the most effective tool for test and troubleshooting of digital equipment (computers, microprocessor-based systems, etc.). Not only can the logic analyzer test many digital circuits, but it can trace and display each step of a digital program. However, an oscilloscope makes an excellent companion to the logic analyzer, although some technicians prefer to use the logic pulser, logic probe, current tracer, logic clip, and logic comparator in place of the oscilloscope.

If an oscilloscope is used in digital equipment testing, the oscilloscope must have good transient response and a wide passband (because of high-speed digital pulses). Also, the usefulness of an oscilloscope in digital equipment testing depends on the features or capabilities of the instrument. For example, an oscilloscope with a multitrace capability permits the two inputs and one output of a typical digital OR circuit to be monitored simultaneously. A dual-trace oscilloscope can monitor the input and output of a flip-flop circuit simultaneously.

In many cases, the oscilloscope is used in a routine manner when testing digital circuits. For example, voltage and current can be measured as described in Chapter 6; time frequency and phase are measured as described in Chapter 7. If certain digital data pulses or "bits" are nonrepetitive, the pulses must be photographed (unless a storage oscilloscope is used). The gain of operational amplifiers used in analog computers can be measured as described in Chapter 10.

In actual practice, the technician, to use an oscilloscope intelligently, must know exactly what kind of signal should appear at each point in a digital circuit. Operation of digital circuits, as well as the highly specialized techniques of digital equipment service, are beyond the scope of this book. However, the following paragraphs describe some of the common applications for oscilloscopes used in digital equipment test. For a thorough discussion of digital equipment circuits, as well as the related test and troubleshooting techniques, your attention is invited to the author's best selling *Handbook of Digital Electronics* (Englewood Cliffs, N.J.: Prentice-Hall, Inc., 1981).

12–16–1.   Monitoring Digital Pulses with an Oscilloscope

Figures 12–20 through 12–29 summarize the most common applications of the oscilloscope in digital equipment testing. In all cases, the internal recurrent sweep is used, and the sweep/sync controls are set to display a suitable number of pulses. Although the amplitude and duration (or frequency) of pulses can be measured with the test connections shown, the relationship of pulses (input to output, memory to clock) is the factor of major interest.

Figure 12–20 shows an oscilloscope used to check the input and output of an inverter circuit. If the oscilloscope has an invert function, the output pulse trace can be inverted (so that both pulses appear positive or negative), and the traces can be superimposed to show a comparison of input to output (in relation to amplitude, duration, and timing). Similarly, if the oscilloscope has an ADD function (both channels added algebraically) the pulses can be superimposed and the ADD function selected. Under these conditions, any pulse trace that remains is the result of a difference between input and output pulses. For example, if the pulses are as shown in Fig. 12–20, and they are added algebraically (pulses superimposed, and the ADD function selected), the input should cancel the output, resulting in a straight horizontal line. If the output pulse is of higher amplitude than the input, there will be partial cancellation, and the trace will be a positive pulse of limited amplitude.

Figure 12–21 shows an oscilloscope used to check the output of a memory circuit in relation to the clock pulses. In digital equipment, it is common for a large number of circuits to be synchronized, or to have specific time relationships to each other. Many of the circuits are frequency dividers, but

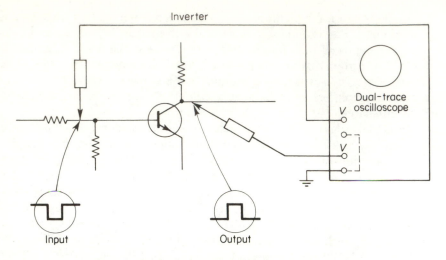

FIGURE 12-20 Checking inverter circuit operation.

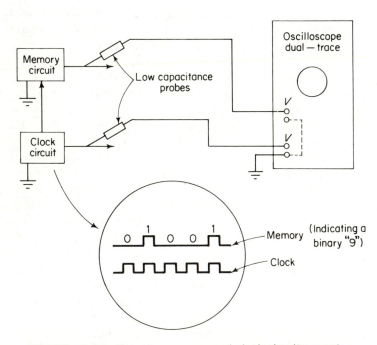

FIGURE 12-21 Checking memory and clock circuit operation.

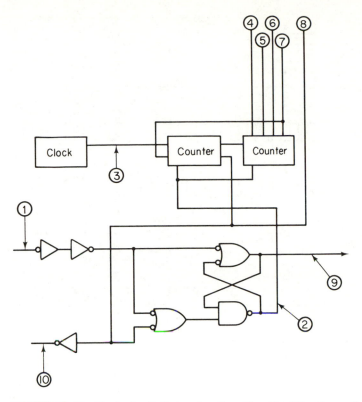

FIGURE 12-22 Typical digital circuit with identification of waveform test points.

waveforms are often time-related in many other combinations. Figure 12–22 shows a typical digital circuit and identifies several of the points at which waveform measurements are appropriate. The accompanying Fig. 12–23 shows the normal waveforms to be expected at each of these points and their timing relationships. The individual waveforms have limited value unless their timing relationship to one or more of the other waveforms is known to be correct. A dual-trace oscilloscope allows this comparison to be made. In typical fashion, waveform 3 is displayed on channel A and waveforms 4 through 10 are successively displayed on channel B, although other timing comparisons can be made.

In the family of time-related waveforms shown in Fig. 12–23, waveform 8 or 10 is an excellent sync source for viewing all the waveforms, since there is but one trigger pulse per frame. For convenience, external sync using waveform 8 or 10 as the sync source may be desirable. With external sync, any of the waveforms may be displayed without readjustment of the sync controls.

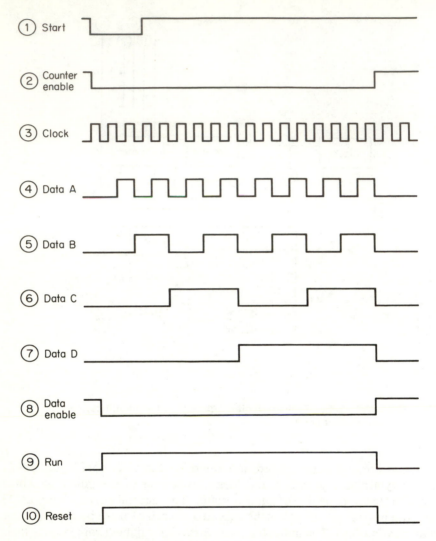

FIGURE 12-23 Typical digital circuit waveforms.

Waveforms 4 through 7 should not be used as the sync source because they do not contain a triggering pulse at the start of the frame. It is not necessary to view the entire waveforms as shown in Fig. 12-23 in all cases. In fact, there are may times when a closer examination of a portion of the waveforms is appropriate. In such cases, it is recommended that the sync remain unchanged while the sweep speed or magnification is used to expand the waveform display.

Figure 12-24 shows an oscilloscope used to check the operation of a

302

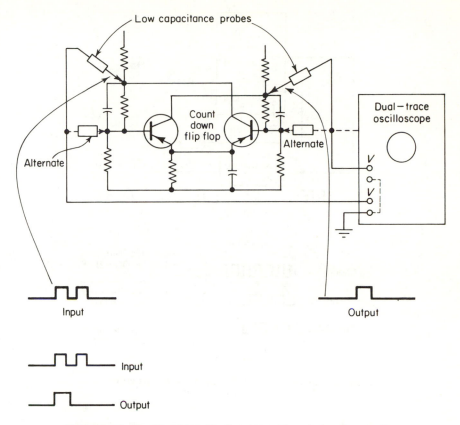

FIGURE 12-24 Checking flip-flop (countdown) circuit operation.

basic flip-flop circuit (used to provide countdown or frequency division). Figure 12-25 shows the waveforms involved in a basic divide-by-two circuit. Figure 12-25a indicates the reference or clock pulse train. Figure 12-25b and c indicate the possible outputs of the divide-by-two circuitry. In Fig. 12-25c, the divide-by-two output waveform is shown for the case where the output circuitry responds to a negative-going waveform. In this case, the output waveform is shifted with respect to the leading edge of the reference frequency pulse by a time interval corresponding to the pulse width.

Figure 12-26 indicates waveform relationships for a basic divide-by-eight circuit. The reference frequency of Fig. 12-26a is supplied to the oscilloscope channel A input. Figure 12-26b shows the *ideal* time relationships between the input and output pulses. In a digital equipment circuit operating at or near there maximum design frequency, the accumulated effects of the consecutive stages produce a built-in *time propagation delay* which can be significant in a critical circuit. Figure 12-26c indicates the possible time delay that

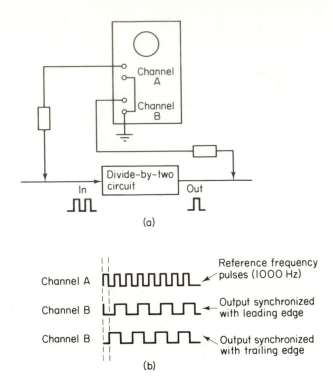

(a)

Channel A — Reference frequency pulses (1000 Hz)

Channel B — Output synchronized with leading edge

Channel B — Output synchronized with trailing edge

(b)

FIGURE 12-25 Checking a basic divide-by-two circuit.

may be introduced into a frequency-divider circuit. By use of a dual-trace oscilloscope, the input and output waveforms can be superimposed to determine the exact amount of propagation delay that occurs.

Figure 12-27 shows a means of measuring propagation time delay when the oscilloscope has both the invert and ADD functions. In Fig. 12-27a, the input to a divide-by-eight circuit is applied to channel A, and the output of the divide-by-eight circuit is applied to channel B, in the normal manner (ADD function not used; neither channel inverted). In Fig. 12-27b, the ADD mode is selected and neither channel is inverted. In Fig. 12-27c, the ADD mode is selected and channel B is inverted. In all three cases, propagation time (T_P) is measured from the leading edge of the ninth reference pulse (channel A). As shown, a more precise measurement can be obtained if the T_P portion of the waveform is expanded or magnified.

Figures 12-28 and 12-29 show oscilloscopes used to check an OR gate and an AND gate, respectively. Note that with an OR gate, the circuit produces an output pulse when either or both input pulses are present. An AND gate produces an output pulse only when there is a coincidence of two input pulses. The same procedures shown in Figs. 12-28 and 12-29 can be applied to

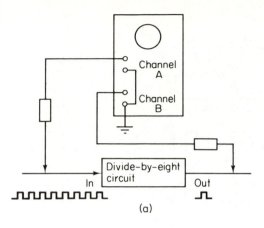

(a)

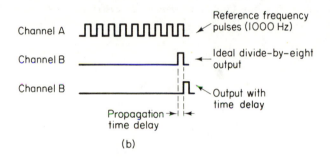

(b)

FIGURE 12-26 Checking a basic divide-by-eight circuit.

tests of other gates used in digital electronic equipment (such as NAND, NOR, EXCLUSIVE-OR, and EXCLUSIVE-NOR).

12-16-2. Troubleshooting Digital Equipment with an Oscilloscope

Present-day digital equipment is often programmed. The classic approach for troubleshooting any programmed device is to monitor a significant system function (such as the data and address buses), go through each step in the program, and compare the results with the program listing for each address and step. This is sometimes known as a *program trace* or, more simply, a *trace* function. One technique for this procedure is called *single stepping*. With the single-stepping approach, you remove the normal clock pulse and replace them with single, one-at-a-time pulses obtained from a switch or pushbutton. This permits you to examine and compare the data at each address with that shown in the program.

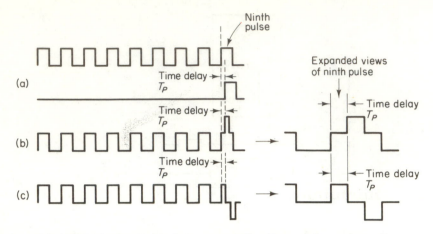

FIGURE 12-27 Measuring propagation-time delay when the oscilloscope has both the invert and ADD functions. (a) Normal; (b) ADD, not inverted; (c) ADD, channel B inverted.

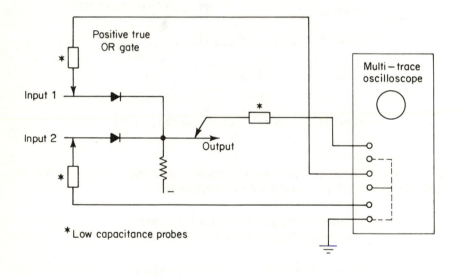

*Low capacitance probes

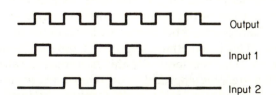

FIGURE 12-28 Checking OR-circuit operation.

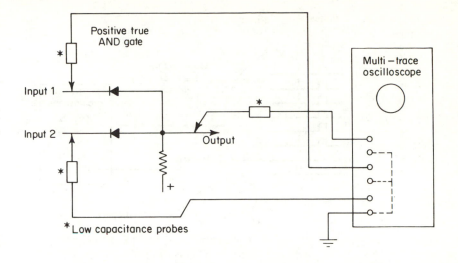

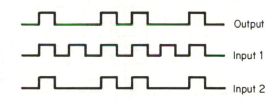

FIGURE 12-29 Checking AND-circuit operation.

For example, assume that each of the eight lines on the data bus are connected to a multitrace oscilloscope so that the bit on each line appears as a pulse on a corresponding trace. A pulse on the trace indicates a binary 1, whereas the absence of a pulse indicates a binary 0 on that line, as shown in Fig. 12–30. The test is started by applying a single pulse to the reset or clear line. Then sufficient single pulses are applied to the clock line until address 1 (0001) appears on the address bus. Now assume that the program listing shows that a hex 7F (binary 0111–1111) data byte should be on the data bus at address 0001 but that the oscilloscope traces show binary 0011–1111. Obviously, the wrong instruction will be applied to the microprocessor and the system will malfunction. This can be caused by a broken line on the data bus, by a defect in the memory, by the absence of a memory-read pulse or a memory-read pulse that appears at the wrong time (opens the memory data buffer too soon or too late), or by several other possible causes. However, you have isolated the problem and determined where it occurs in the program.

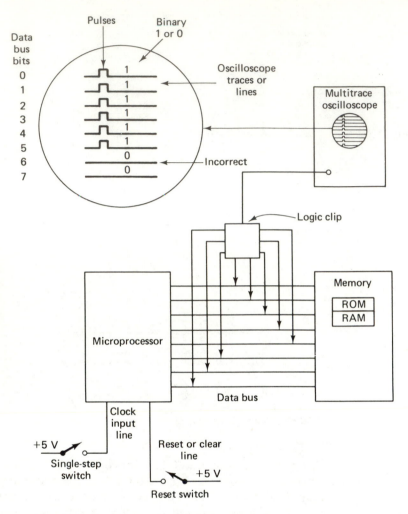

FIGURE 12-30 Basic microprocessor troubleshooting approach using multitrace oscilloscope and single stepping.

If the problem appears to be one of *timing,* the oscilloscope can be used to check the time relationship of the related pulses. For example, the oscilloscope can be connected to the data bus, address bus, and read line as shown in Fig. 12-31. The oscilloscope then shows the time relationships among the pulses on these lines.

In this simplified example, the read pulse must hold the memory data buffer closed until the selected address pulses appear on the address bus (sometimes known as the *valid address* point), must hold the buffer open just

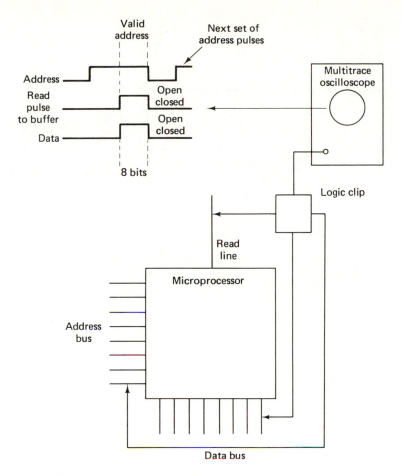

FIGURE 12-31 Basic microcomputer troubleshooting approach for timing problems.

long enough for all 8 data bits to appear on the data bus, and then must close the buffer until the next address is applied. In a practical case, the entire timing diagram (such as shown in Fig. 12–23) can be duplicated on a multitrace oscilloscope.

13

Checking
Television Receivers

An oscilloscope sweep generator/marker generator is the best instrument for testing a black-and-white TV receiver. A color TV receiver requires a color generator. This chapter describes the basic alignment procedures for a black-and-white TV receiver using the oscilloscope/sweep generator combination. Remember that a color TV receiver is essentially a black-and-white set, except for the color display circuits. Therefore, the basic alignment procedures are the same.

13-1. CHECKING OPERATING WAVEFORMS

An oscilloscope is the best instrument for checking the amplitude, frequency, and waveshape of video and sync pulses in any TV receiver. These waveshapes can be checked without a signal generator of any type, since many of them are generated in the TV receiver itself or are produced by the incoming transmitted TV signal.

1. Connect the equipment as shown in Fig. 13–1. Have a low-capacitance probe and a demodulator probe available.

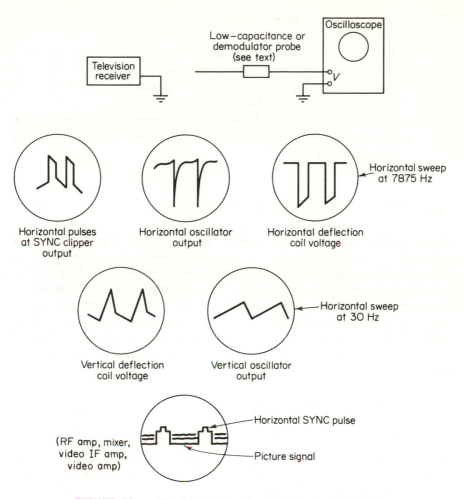

FIGURE 13-1 Checking TV receiver operating waveforms.

2. Place the oscilloscope in operation (Chapter 5). Switch on the internal recurrent sweep. For vertical signals, set the sweep frequency and sync controls to 30 Hz. This will display two complete cycles of the waveform. For horizontal signals, set the sweep and sync controls for 7875 Hz. This will also display two complete cycles of the waveform.

NOTE

Many oscilloscopes designed especially for TV service have preset horizontal sweep frequency positions of 30 and 7875 Hz.

3. Switch on the TV receiver, tune in the desired channel, set the con-

trast, hold, and other receiver controls to their normal operating position.

4. In turn, connect the oscilloscope probe to each point of interest in the TV receiver. Use a demodulator probe for signals in any stage ahead of the video detector. Use a low-capacitance probe for signals in and after the video detector.

5. Check the waveforms against those of Fig. 13-1 or, preferably, against the manufacturer's service data. The waveforms of Fig. 13-1 are typical.

13-2. BASIC TV ALIGNMENT PROCEDURE

The general procedure for alignment of split-sound and intercarrier types of television receivers is the same, the major differences being in the number of intermediate frequencies used and the frequencies employed. There are four separate steps for overall alignment: tuner or RF alignment, picture (video) IF alignment, trap alignment, and sound IF and FM detector alignment. The actual procedures for these alignments should be followed as described in the manufacturer's service data. The actual response waveforms should be checked against those shown in the service data, using the following basic procedures as a guide.

13-2-1. Tuner Alignment

If the tuner is defective, it is generally better to work with the tuner set to only one channel position until the trouble is corrected. Then, other channel positions can be compared with the initial channel for sensitivity, switching noise, and general performance.

If the tuner is satisfactory in these respects, it is advisable to check the alignment by observing the response curves for each channel. Curves for individual channels should be examined and compared with those shown in the manufacturer's service notes. If a response curve indicates that alignment is required, the technician should refer to the alignment curves given in the service data as guides and follow closely the recommended alignment procedure. Alignment should not be attempted until these preliminary tests have been completed. In principle, complete front-end alignment includes alignment of the antenna input circuits and adjustment of the amplifier and RF oscillator circuits. Most tuners merely require "touch up," in which relatively few of the adjustments are used.

Either way, the adjustments require that a sweep signal and marker signal be fed into the tuner so that a response curve with markers will be reproduced on the oscilloscope screen. Alignment is accomplished by setting adjustments so that the waveshape on the oscilloscope screen resembles the waveshape shown by the manufacturer in the service notes. Signals from the marker generator are used to provide frequency reference points in shaping the curve.

Serious misalignment of the tuner or considerable difficulty or failure in alignment may indicate a defective component. If proper alignment procedure fails to produce correct tuner curves, the technician should check individual components in the RF unit.

The RF tuner can be aligned using either the direct-signal injection method, or with the aid of marker adder units.

NOTE

With the marker adder system of alignment, also known as *post-injection* or *bypass alignment,* the marker signal is added to the sweep response curve by the marker adder unit after the demodulated sweep signal is taken out of the receiver under test. With conventional direct-signal injection method, in which the sweep and marker signals are both passed through the receiver circuits, overloading or clipping by the receiver circuits can introduce distortion of the marker and distortion of the sweep curve by the marker. The marker adder system of adding markers to the response curve eliminates this source of distortion. In addition, the marker adder system permits simple and precise alignment of a variety of trap circuits without marker "suckout."

1. Connect the equipment as shown in Fig. 13–2 for marker adder alignment, or as shown in Fig. 13–3 for direct-signal injection.

2. Place the oscilloscope in operation (Chapter 5). Switch off the internal recurrent sweep. Set the sweep selector and sync selector so that the horizontal sweep is obtained from the generator sweep output. If the generator sweep is at line frequency, the oscilloscope horizontal sweep can be set to line. The phasing control on the oscilloscope or sweep generator must be set to produce a single pattern (refer to Chapter 8).

3. Switch on the sweep generator and marker generator as described in their respective instruction manuals.

4. Disable the AVC-AGC circuits in the receiver under test. Use the bias recommended by the manufacturer's service notes. Usually, this is about − 1.5 V. Make certain to apply the negative bias to the AVC-AGC line, and the positive bias terminal to the receiver chassis.

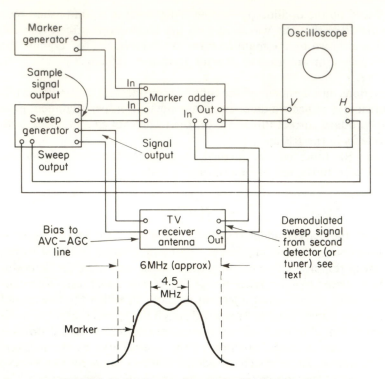

FIGURE 13-2 Tuner alignment with marker adder.

5. For marker adder alignment, connect the demodulated sweep pick-up lead to the receiver video output.

NOTE

This test point may be across the load resistor of the receiver second detector if an "overall" response curve (RF, IF, detector) is desired. If the shape of the tuner curve is to be checked, connect the lead to the test point (known as the "looker point") in the tuner mixer circuit. On some tuners, a demodulated signal can be obtained at this point. However, the oscilloscope should be connected to the point through an isolating resistor (about 50 kilohms). On other tuners, it will be necessary to demodulate the sweep signal by means of an RF probe. In these cases, the manufacturer's service notes should be consulted for the correct procedure.

6. For direct-injection alignment connect the oscilloscope probe to the receiver second detector.

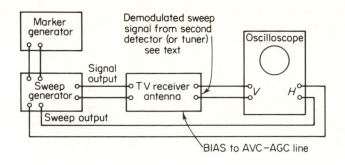

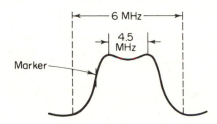

FIGURE 13-3 Tuner alignment sweep and marker by direct injection.

7. Adjust the sweep and marker generators to the appropriate channel frequency. Set the sweep width to at least 10 MHz. Adjust the oscilloscope horizontal and vertical gain controls as necessary for a trace similar to that of Fig. 13–2 or 13–3.

13-2-2. Picture (Video) IF Alignment

If a television receiver is to give wideband amplification to the television signal, the picture IF (video) system of the receiver must pass a frequency band of approximately 4 MHz. This is necessary to ensure that all the video information is fed through to the picture tube grid, and the resultant picture has full definition. The bandpass of color TV receivers must be essentially flat to beyond 4 MHz to ensure that color information contained in the color sidebands is not lost.

The IF stages can also be checked and aligned, using either the direct-signal injection method, or marker adder units (postinjection or bypass alignment).

1. Connect the equipment as shown in Fig. 13–4 for marker adder alignment, or as shown in Fig. 13–5 for direct-signal injection.
2. Place the oscilloscope in operation (Chapter 5). Switch off the internal recurrent sweep. Set the sweep selector and sync selector so that

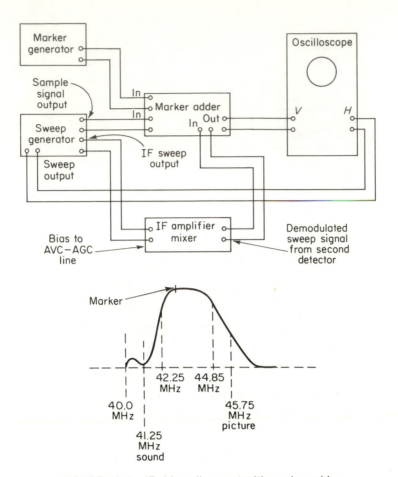

FIGURE 13-4 IF video alignment with marker adder.

the horizontal sweep is obtained from the generator sweep output. If the generator sweep is at line frequency, the oscilloscope horizontal sweep can be set to line. The phasing control on the oscilloscope or sweep generator must be set to produce a single pattern (refer to Chapter 8).

3. Switch on the sweep generator and marker generator as described in their instruction manuals.

4. Disable the AVC-AGC circuits in the receiver under test. Use the bias recommended by the manufacturer's service notes. Usually, this is about −1.5 V. Make certain to apply the negative bias to the AVC-AGC line, and the positive bias terminal to the receiver chassis.

5. For marker adder alignment, connect the marker adder demodu-

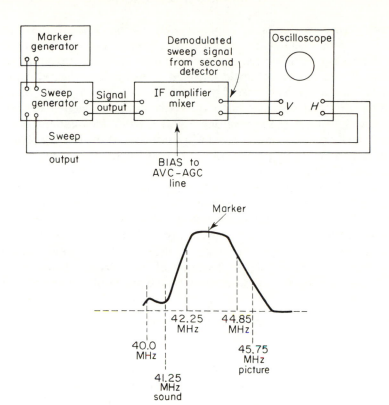

FIGURE 13-5 Video IF alignment by direct sweep and marker injection.

lated sweep pickup lead to the receiver video output (second detector). For direct-injection alignment, connect the oscilloscope probe to the second detector.

6. Adjust the sweep and marker generators to the appropriate intermediate frequency. Set the sweep width to at least 5 MHz. Adjust the oscilloscope horizontal and vertical gain controls as necessary for a trace similar to that of Fig. 13-4 or 13-5.

13-2-3. IF Sweep Response Curve Analysis

Examples of IF response curves are shown in Figs. 13-4 and 13-5. The frequency relation of the sound carrier to the picture carrier is reversed in the IF amplifier (from that of the RF tuner) because the receiver local oscillator operates at a frequency higher than that of the transmitter carrier.

Note the following two characteristics of the picture IF response curve:

(1) the picture carrier is at approximately 40 or 50% of the maximum response; (2) the sound carrier must be at 1% or less of maximum response. The picture carrier is placed at approximately 40% of maximum response because of the nature of single-sideband transmission, the system used in transmitting television signals. If the circuit is adjusted to put the picture carrier too high on the response curve, the effect will be a general decrease in picture quality caused by the resulting low-frequency attenuation; placing the picture carrier too low on the curve will cause loss of the low-frequency video response and result in poor definition. Loss of blanking and proper synchronization will also occur.

The skirt selectivity of the picture IF curve is made sharp enough to reject the sound component of the composite signal. The sound carrier is kept at a low level to prevent interference with the video signal. To achieve this selectivity in split-sound receivers, an absorption circuit, consisting of a trap tuned to the sound intermediate frequency, is used. Some receivers include additional traps tuned to the higher frequency of the adjacent channel sound carrier. These traps have a marked effect on the shape of the response curve.

13-2-4. Alignment of Sound IF Amplifiers and FM Detectors

The procedure used in aligning sound IF amplifiers in both TV and FM receivers is similar. Intercarrier-type television receivers use a sound IF of 4.5 MHz. Split-sound-type receivers may use either 21.25 or 41.25 MHz. FM receivers use 10.7 MHz. Procedures for overall alignment of the sound section, including both the IF amplifier and FM detector, may vary, depending upon the type of FM detector. The general procedures of Secs. 11–6 and 11–8 should apply to most television receivers.

13-2-5. Trap Alignment

One or more traps may be contained in the RF tuner and picture and sound IF amplifiers, depending upon the type of television receiver. Traps are included to attenuate specific frequencies, such as adjacent picture and sound carriers, or picture and sound IF signals in various parts of the receiver.

The test equipment setup for alignment of traps in various stages is essentially the same as for alignment of the particular stages (refer to Secs. 13–2–1 through 13–2–4). However, certain problems may be encountered.

Because the response of the amplifier is very low at the trap frequencies, the marker may often be difficult to see on the oscilloscope response curve. The use of marker adders is recommended for trap alignment. If extreme difficulty is encountered, the traps can be set with a d-c voltmeter rather than an oscilloscope. With such an arrangement, the d-c voltmeter is connected across the sound detector load resistor, the marker generator is set for the trap frequency, and the trap is tuned for a minimum voltage reading on the meter.

The general procedure in aligning picture IF amplifiers is first to set the traps, and then to align the amplifier circuits. Since any adjustment of the amplifier circuits will usually slightly detune the traps, the traps may have to be "touched up" during the picture IF amplifier alignment. Again, the manufacturer's alignment instructions determine the exact procedure to follow.

13-3. TVV AND TVH SYNC AND SWEEPS

As discussed in Sec. 13-1, many oscilloscopes designed specifically for TV service have preset horizontal sweep frequency functions of 30 and 7875 Hz, called the TVV and TVH functions. These sweeps are usually selected by means of oscilloscope controls, and permit pairs of vertical or horizontal sync pulses to be displayed on the oscilloscope screen. As shown in Fig. 13-6, when TVV is selected, the oscilloscope horizontal sweep is set to a rate of 30 Hz, and the horizontal trigger is taken from the TVV output of a sync separator in the oscilloscope.

The oscilloscope sync separator receives both 60-Hz and 15,750-Hz (vertical and horizontal) sync pulses from the TV receiver signals being moni-

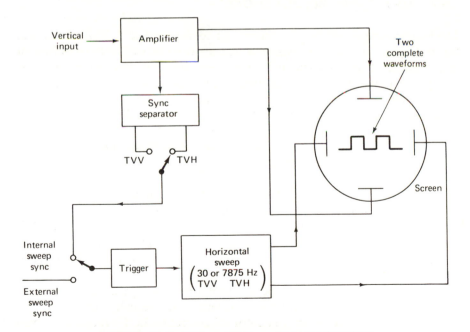

FIGURE 13-6 TVV and TVH oscilloscope functions.

tored. The oscilloscope separator operates similarly to the receiver sync separator and delivers two separate outputs (if both sync inputs are present). If only one sync input is present, only one output is available. Either way, the selected output (60 or 15,750 Hz) is used to synchronize the horizontal sweep trigger.

With the horizontal sweep at 30 Hz and the trigger at 60 Hz, there are two pulses for each sweep, and two vertical sync pulses (or two complete vertical displays) appear on the screen. When TVH is selected, the horizontal sweep is set to a rate of 7875 Hz, and the trigger (15,750 Hz) is taken from the horizontal output of the oscilloscope's sync separator. Again, two pulses or displays are presented on the screen since the sweep is one-half the trigger rate.

The TVV and TVH features not only simplify observation of waveforms (you always get exactly two complete displays), but they can also be helpful in trouble localization. As an example, assume that you are monitoring waveforms in the TV receiver at some point where both vertical and horizontal sync signals are supposed to be present. (This can be anywhere ahead of the TV receiver sync separator.) If you find two steady patterns on the TVV position but not in the TVH position, you know that there is a problem in the horizontal circuits or that horizontal sync pulses are not getting to the point being monitored.

13-4. USING THE VECTORSCOPE

The basic principles of the vectorscope are covered in Sec. 1–8. As discussed, the vectorscope is used in television broadcast work as well as to troubleshoot and adjust the color demodulators of TV receivers. The vectorscope also permits you to judge the general condition of all color circuits (chroma bandpass, burst amplifier, reference oscillator, phase detectors, etc.). In this section, we cover basic vectorscope techniques that apply to practical television service.

13-4-1. Connecting a Conventional Oscilloscope as a Vectorscope

1.   Connect the oscilloscope vertical input to the driven element of the red gun, usually the grid, as shown in Fig. 1–11. If the cathode is the driven element, then connect the vertical probe tip to cathode. (The driven element is the element to which the output signal of the color amplifier is applied.) Use a low-capacitance probe. Connect the probe ground lead to the TV receiver chassis.

2.   Connect the oscilloscope horizontal input to the driven element of the blue gun. Connect the probe ground lead to the receiver chassis.

3.   Connect a color-bar generator RF output to the TV receiver antenna

terminals. Reset the channel selector, fine tuning, and other controls as necessary to obtain a color-bar pattern on the TV receiver screen. Turn off any special color controls, such as "accutint," automatic color control, and so on.

4. Reset the oscilloscope controls as necessary to obtain a circular rosette pattern on the oscilloscope, as shown in Fig. 1–9. Note that the pattern has 10 petals or vectors, one for each color bar. The third petal (counting clockwise from the upper left area of the trace) represents the red bar. With proper adjustment of the hue or tint control, the third bar should be located at or near the top of the pattern (that is, near a 12 o'clock position).

5. If an overlay (such as shown in Fig. 1–10) is available for the oscilloscope, attach the overlay and use the reference marks to align the petals.

13–4–2. *Basic Vectorscope Alignment Procedure*

1. Turn the tint or hue control through its range, and note the effect on the vector pattern. The pattern will turn but should *not change* in size. If necessary, adjust the 3.58-MHz oscillator coil (of the TV receiver) until the pattern is the same size throughout the range of the hue or tint control.

2. Set the tint or hue control to the center of its range. Adjust the reactance coil (or oscillator trimmer capacitor in some receivers) so that the color signal stays in sync as the chroma control on the color-bar generator is turned to minimum. The vector pattern will reduce in size, but it *should not* rotate. Rapid spinning of the vector pattern indicates misadjustment of the reactance coil (or oscillator trimmer).

3. Set the tint or hue control the center of its range. Adjust the burst phase transformer (TV receiver) to move the third vector (corresponding to the red bar) to the 12 o'clock position. Readjust the reactance coil (or trimmer capacitor) simultaneously to maintain color sync (no rotation of vector pattern).

4. Rotate the tint control fully in each direction from the center range position, and note the effect on the vector pattern. If necessary, readjust the burst phase transformer so that the third vector moves an *equal distance* each side of the normal vertical position as the tint control is varied in each direction from midrange. The tint control should have sufficient range to rotate the pattern so that the third vector moves to each side at least as far as the normal position of the adjacent bars (vectors 2 and 4). In many receivers, it is possible to rotate the red vector as much as 45° in each direction.

5. Repeat steps 2 through 4 until best results are obtained. If proper adjustment cannot be made, check to be sure that the 3.58-MHz coil cores are set to their proper peak. Also refer to the service literature for information on color circuit alignment.

13-4-3.   Interpreting Vectorscope Patterns

The display of Fig. 13-7 is that produced by color receivers using a 90° demodulator system (which is typical of the Zenith color system). In a typical vectorscope (or on a vectorscope overlay), the reference burst signal is considered as 0° (or no phase shift). This is represented by the 0, or burst mark, on the vectorscope screen or overlay. The red bar of the standard 10-bar pattern is the third bar and is 90° from the burst. This is at the top of the vectorscope screen or overlay. The red bar is marked R-Y. The numbers 1 and 2 represent the first and second bars of the pattern and are all 30° apart on the screen. The B-Y mark (blue) is 90° away from the R-Y mark and 180° from the burst.

13-4-4.   Basic Demodulator Adjustments with Vectorscope Patterns

The vectroscope pattern is very useful for adjusting the phase angle between the red and blue demodulators for a 90°, 105°, or 116° demodulation system. (Note that the demodulators of some receivers are not adjustable.)

To align a 90° demodulator system (Zenith type, Fig. 13-7), adjust the demodulators until the third bar rests on the R-Y mark and the sixth bar is on the B-Y mark.

If the receiver calls for a demodulator angle of 105° (typical of the General Electric type, Fig. 13-8), the third bar should rest on the R-Y mark, with the sixth bar halfway between the B-Y and No. 7. Note that the pattern of Fig. 13-8 is not as round as that of Fig. 13-7. This is due to the increased angle of demodulation and is normal (for that type of demodulator).

As the angle is increased to 116° (such as the old Motorola-type demodulation, Fig. 13-9), the pattern appears more squared, as shown. The sixth bar is past the No. 7 mark.

As can be seen, the overall shape of vector patterns describes the type of demodulation used as well as the existing state of demodulator adjustment.

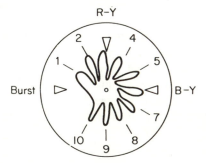

FIGURE 13-7 Vectorscope display produced by color receivers using a 90° demodulator system.

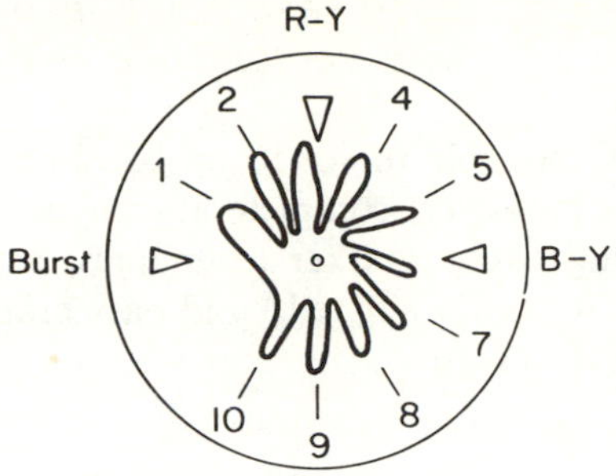

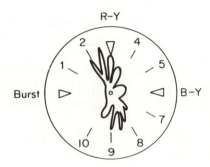

FIGURE 13-8 Vectorscope display produced by color receivers using a 105° demodulator system.

FIGURE 13-9 Vectorscope display produced by color receivers using a 116° demodulator system.

Always consult the service literature concerning recommended demodulator alignment procedures (and corresponding vector patterns).

13-4-5. Alignment of an Off-Frequency 3.58-MHz Oscillator

The following procedure should be used only when the 3.58-MHz oscillator is way off frequency.

1. Disable the correction signal to the 3.58-MHz oscillator as described in the service literature. Usually, the correction signal can be disabled by a bias applied at some test point in the circuit.

2. With the corresion signal disabled, the oscillator is "free-wheeling," and the pattern appears to rotate or possibly appears as a blurred circle, depending on how far the oscillator is off frequency. (A free-running oscillator can also be verified by a "barber-pole" effect on the receiver screen. The color bars are in a broken diagonal pattern.)

3. Adjust the 3.58-MHz oscillator until the vectorscope pattern stands still, or is as close as possible to a motionless condition. Then restore the correction signal to the oscillator.

13-4-6. Band-Pass Amplifier Alignment with Vectorscope Patterns

It is possible to touch up band-pass amplifier alignment using a vectorscope. However, complete alignment of any stage in a receiver should be done using sweep/marker techniques, as described in other sections of this chapter. A poorly aligned band-pass amplifier produces a display similar to that of Fig. 13-10.

13-4-7. Troubleshooting with Vectorscope Patterns

A loss of the R-Y signal causes a lack of vertical deflection and the pattern appears as shown in Fig. 13-11. The B-Y signal deflects the beam along the horizontal axis, producing a bright line. This indicates that the trouble lies in the R-Y demodulator, matrix, or difference amplifier, depending on the circuit used in the receiver. If the R-Y signal is weak, some deflection will be noted, and an extremely distorted pattern results, again pointing to the R-Y circuits.

A loss of the B-Y signal results in a lack of horizontal deflection, and produces a pattern similar to that of Fig. 13-12. The R-Y signal causes the beam to deflect vertically. This indicates that the problem is in the B-Y dif-

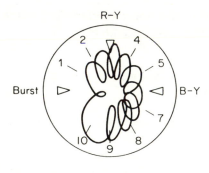

FIGURE 13-10 Vectorscope display produced by color receiver with poorly aligned band-pass amplifier.

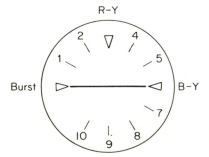

FIGURE 13-11 Vectorscope display produced by color receiver with loss of R-Y signal; no vertical deflection.

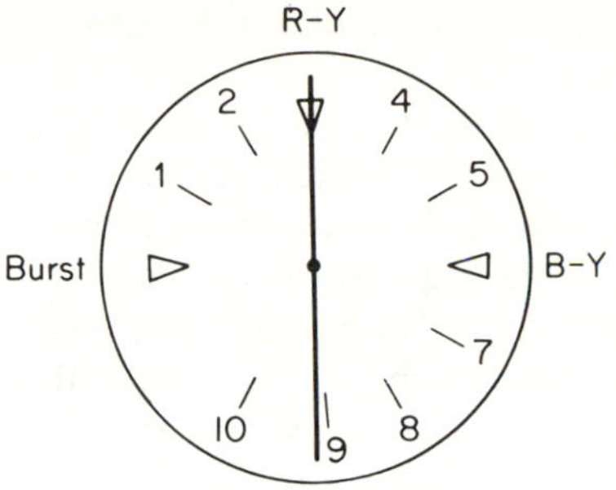

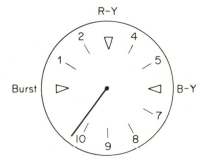

R-Y

FIGURE 13-12 Vectorscope display produced by color receiver with loss of B-Y signal; no horizontal deflection.

Burst

B-Y

R-Y

FIGURE 13-13 Basic connections for ringing test of coils, transformers, and yokes.

Burst

B-Y

ference amplifier, matrix, or demodulator circuits. Again, if the B-Y is weak, some deflection will be noted, and a distorted display results.

If there is a complete loss of color, the pattern will appear somewhat like that in Fig. 13-13.

Keep in mind that not all circuits produce identical vectorscope patterns. For example, as demodulation angle is increased (such as with General Electric and Motorola), the pattern appears more "square" than with 90° demodulation systems (such as Zenith), even though all the receivers are operating properly.

The patterns discussed here are for reference only and must not be considered as typical. Always consult the vectorscope instruction manual and all service literature.

13-5. TESTING TRANSFORMERS, YOKES, AND COILS

Transformers, yokes, and coils are often difficult components to test. Of course, if one of these components has a winding that is completely open or shorted, or that has a very high resistance, this shows up as an abnormal voltage and/or resistance indication during troubleshooting. However, a winding that has a few shorted turns or is leaking to another winding may pro-

325

duce voltage and resistance indications that appear close to normal and pass unnoticed.

An oscilloscope can be used to perform a *ringing* test on such components. In this procedure, a pulse is applied to the component under test. The condition of the component is then evaluated by the amount of *damping* observed in the waveform

The best way to obtain reliable results from a ringing test is to compare the ringing waveform obtained from the part being tested with the waveform from a duplicate part that is known to be good. However, there are many times when a duplicate part is not available for comparison, and you must judge the part being tested by a study of the waveform. To gain experience in evaluating the ringing waveform, it is helpful to try the procedure several times, both with good parts and with parts that you have purposely shorted with various resistances. Although the following procedures apply specifically to television coils and transformers, the procedures can also be used to test any component that has one or more windings.

13-5-1. Basic Ringing Test Procedure

Connections for the basic ringing test are shown in Fig. 13–14. The basic ringing test procedure is as follows:

1.   Remove power from the part or circuit to be tested. Do not apply power to the circuit at any time during the test procedure.
2.   Disconnect the part to be tested from the related circuit. Although many parts can be tested in-circuit, it is usually necessary to remove circuit connections, especially in solid-state circuits where the diodes and transistors have a loading effect on the part.

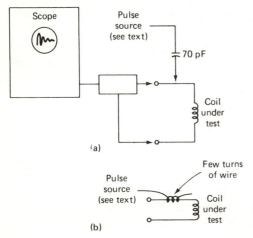

FIGURE 13-14 Basic connections for ringing test of coils, transformers, and yokes.

3.   Connect the oscilloscope to the part terminals as shown. Use a low-capacitance probe.

4.   Connect a pulse source to the part through a capacitor as shown in Fig. 13–14a, or through a few turns of wire as shown in Fig. 13–14b. All oscilloscopes have circuits that produce pulses suitable for using in ringing tests. (These pulses are the same ones used to trigger the oscilloscope horizontal sweep.) Unfortunately, the pulses are not always accessible without modification of the oscilloscope. However, many oscilloscopes designed specifically for TV service have an external terminal (either front panel or rear panel) that provides for electrical connections to the pulse source.

5.   Operate the oscilloscope controls to produce a ringing waveform as shown in Fig. 13–15a or b. The waveform of Fig. 13–15a is a typical ringing pattern for a normal coil or transformer winding. The waveform of Fig. 13–15b is representative of patterns obtained from defective coils and transformers (with shorted turns or leakage in the windings). Keep in mind that these waveforms are typical only. Experience is necessary to judge the waveforms when no duplicate parts are available for comparison.

6.   To help you make the judgment, connect a resistance across the part terminals (or make a loop consisting of a few turns of solder and pass the loop around the part). Note any change in the ringing pattern. There should be a drastic change in the pattern when the resistor (or solder coil) is added. If not, the part is suspect.

13-5-2.   Ringing Test for High-Voltage (Flyback) Transformer

To test the high-voltage transformer in the horizontal system of a TV receiver, remove the horizontal output and high-voltage rectifier tubes, or disconnect the corresponding leads in a solid-state receiver circuit. Using the procedure described in Sec. 13–5–1, obtain a ringing waveform across the transformer primary. Typical test connections are shown in Fig. 13–16.

With the ringing pattern established, connect a short across the

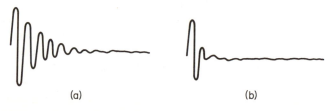

(a)   (b)

FIGURE 13-15   Typical ringing waveforms. (a) Ringing pattern for normal coil or transformer; (b) ringing pattern for defective coils and transformer windings.

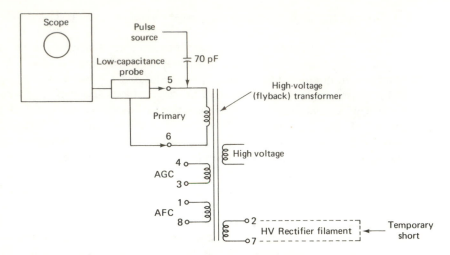

FIGURE 13-16   Ringing test for high-voltage (flyback transformer).

transformer filament winding. In the transformer of Fig. 13–16, connect the short across terminals 2 and 7. If a significant change in the waveform is noted, the transformer is probably good. Little or no waveform change indicates that the transformer is defective.

13-5-3.   Ringing Test for Picture Tube Yoke

To test the horizontal and vertical windings of the deflection yoke, disconnect the yoke from the circuit. Using the procedure described in Sec. 13–5–1, obtain a ringing waveform across each winding, in turn. Typical test connections are shown in Fig. 13–17.

Note that the horizontal winding of the yoke in Fig. 13–17 consists of two sections. Alternately short each section. If the winding is good, the effect on the waveform should be the same as each section is shorted.

The vertical winding of the yoke also consists of two sections, with a damping resistor connected across each section. Disconnect these damping resistors, as well as the capacitor that is in parallel across both vertical sections. Then alternately short each section. If the winding is good, the effect on the waveform should be the same as each section is shorted.

In some yokes, the two sections of each winding are connected in parallel, as shown in Fig. 13–18. In such cases, disconnect the sections at one end. Connect the oscilloscope across both sections, as shown in Fig. 13–18b, and then short each section, in turn. Again, the effect on the waveform should the same as each section is shorted.

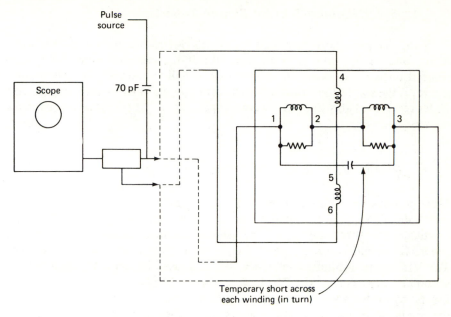

FIGURE 13-17 Ringing test for picture tube yoke (series winding).

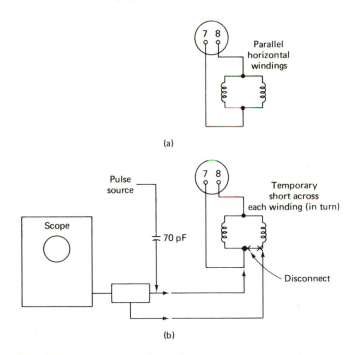

FIGURE 13-18 Ringing test for picture tube yoke (parallel winding).

13-5-4. Ringing Test for Picture Tube Coils

Use the procedure described in Sec. 13–5–1 to test the width, linearity, focus, and so on, coils of a picture tube if they are suspected of having shorted turns or leakage. The solder loop usually produces the best results. Again, if the solder loop causes a significant change in the waveform, the coil under test is probably good. However, experience is the best judge when making any ringing test.

13-6. USING VITS AND VIR SIGNALS

Most network television transmissions contain built-in test signals that can be very valuable tools in troubleshooting and servicing TV receivers. The VITS (vertical interval test signal) can be used to localize trouble to the antenna, tuner, IF, or video sections, and shows when realignment may be required. The VIR (vertical interval reference) signal is used in some TV receivers for automatic color correction. The basic purpose of the VITS and VIR signals is for use by the broadcast engineers, who monitor the test signals to be sure that the quality of their transmission meets established standards. The following procedures show how the TV service technician can analyze and interpret oscilloscope displays of the VITS to troubleshoot TV receivers.

The VITS is transmitted during the vertical blanking interval. On the TV receiver, the VITS can be seen as a bright white line (and/or series of dots) above the top of the picture. However, the vertical hold, linearity, or height controls may require adjustment to view the vertical blanking interval. On some receivers with an internal vertical retrace blanking circuit, the vertical blanking pulse must be disabled to see the VITS. The procedures are covered in the following paragraphs. On other receivers, without internal vertical blanking, it is only necessary to advance the brightness control. For troubleshooting purposes, it is not necessary to see the VITS on the picture tube screen. Instead, you monitor the VITS signals on an oscilloscope after they pass through the receiver.

13-6-1. Composition of the VITS and VIR Patterns

The transmitted VITS and VIR are precision sequences of specific frequencies, amplitudes, and waveshapes, as shown in Figs. 13–19 and 13–20. Keep in mind that the makeup of the VITS varies from one broadcast station to another, so the patterns shown in Figs. 13–19 and 13–20 must be considered as typical only.

In most cases, the VITS contains a "multiburst" signal which begins with a "flag" of white video, followed by sine-wave frequencies of 0.5, 1.5, 2.0, 3.0, 3.6 (3.58), and 4.2 MHz. The multiburst may appear on line 17 or 18,

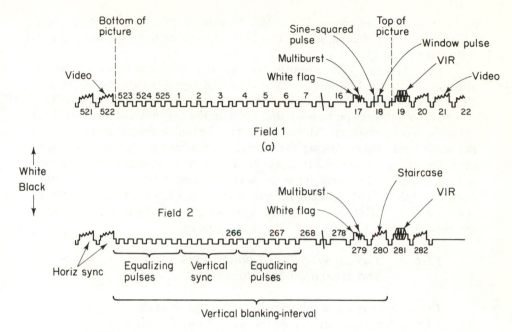

FIGURE 13-19   Vertical blanking interval showing typical VITS and VIR information.

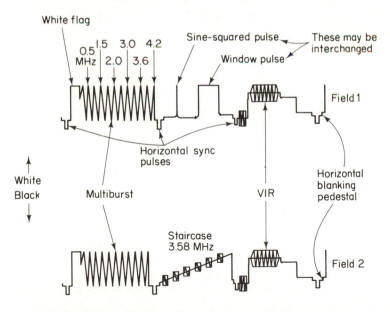

FIGURE 13-20   Typical VITS and VIR signals, fields 1 and 2.

and may be repeated on line 279 or 280. As seen on the television screen, field 1 is interlaced with field 2, so that line 17 is followed by line 279, and line 18 is followed by line 280.

The VITS and VIR appear at the end of the vertical blanking interval and just before the first line of video. The VIR signal is transmitted on lines 19 of fields 1 and 2 (line 19 of field 2 is also identified as frame line 281).

The multiburst portion of the VITS is the portion that can be most valuable to the technician. All frequencies of the multiburst are transmitted at the same level. By examining the comparative strengths of the frequencies after the signal is processed through the TV receiver, you can check the frequency response of the receiver at the normally used video frequencies. Other portions of the VITS, including the sine-squared pulse, window pulse, and the staircase of 3.5-MHz bursts at progressively lighter shading, are variable to the networks, but have less value to the technician.

13-6-2. *Typical VITS Test Connections and Measurement Procedures*

The procedures for monitoring the VITS signals as they are processed by the receiver will vary, depending on the receiver circuits. However, the following steps can be modified as necessary to accommodate all types of TV receivers.

1.   Connect the oscilloscope probe to the output of the video detector or the desired point in the video section of the receiver. Start with the video detector output.

2.   Set the oscilloscope sweep controls for a triggered sweep.

3.   If the receiver has a vertical retrace blanking circuit, the circuit must be disabled. Figure 13–21 shows a typical picture tube circuit with vertical retrace blanking pulses applied to the control grid. These pulses can be bypassed to ground by placing a large capacitance across capacitor C_8 as

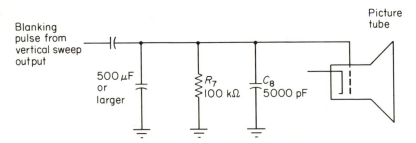

FIGURE 13-21   Disabling the vertical retrace blanking pulses to the picture tube.

shown. On receivers without vertical retrace blanking, simply advance the brightness control.

4. Operate the oscilloscope controls as necessary to obtain a VITS pattern.

5. In typical operation, you must advance the oscilloscope intensity control full on. Then adjust the sweep sync control slowly to the point where triggering barely starts and a horizontal pattern is formed. Make a very slight adjustment of the sweep sync control until the VITS pattern is obtained. Because of the constantly varying video signal, sync lock-in of the VITS can be quite critical. (On some oscilloscopes, additional fine adjustment of the sync level can be obtained by slight readjustment of the vertical gain control.) It is normal for strong video signals to override the VITS pattern occasionally. This effect varies with different TV scenes and may be more noticeable on certain channels. However, with careful adjustment of controls, a good, constant VITS pattern can usually be maintained. On some receivers, you can get more solid lock-in of the VITS pattern by using external sync, as described in step 8.

6. Using the oscilloscope sweep and horizontal gain controls, you can select the portion of the VITS pattern you want to view and then expand it as desired. Generally, you are interested only in the multiburst. As shown in Figs. 13–19 and 13–20, the multiburst portions of the VITS are contained on only 2 horizontal lines of the 525-line raster and that the "staircase" and other portions are on only 1 line. The VITS appears substantially dimmer than the rest of the composite video pattern. The bright vertical line pattern may be distracting. This problem can be minimized by adjusting the horizontal position control to place the bright area just beyond the right side of the screen edge.

7. As alternative, you can display the VITS along with vertical blanking pulses. To produce such a display, set the oscilloscope sweep controls as you would to monitor the vertical composite video waveform. Usually, this involves setting the oscilloscope sweep control to vertical or TVV, or to a similar position. Then you operate the oscilloscope sweep expansion controls to obtain the desired display.

8. As another alternative, you can display the VITS alone (as with a triggered sweep, steps 1 through 6) but with an external sync to lock in the VITS. On some receivers, improved sync stability can be obtained by using external sync, as follows. Locate a point in the receiver where a negative-going (or positive if the receiver has positive sync pulses) vertical sync pulse is available. One convenient point on most receivers is the appropriate wire in the convergence board that leads to the vertical winding of the deflection yoke. Be sure that the receiver vertical and horizontal hold controls are set to provide a stable, in-sync picture. Connect the vertical pulse from the receiver to the external sync input of the oscilloscope. Operate the oscilloscope controls to lock in the VITS pattern and proceed as described in steps 1 through 6.

13-6-3.   Analyzing the VITS Waveform

All frequencies of the multiburst are transmitted at the same level but are not equally passed through the receiver circuits due to the response curve of those circuits (RF, IF, and video). Figure 13–22 shows the desired response for an "ideal" color receiver, identifying each frequency of the multiburst and showing the allowable amount of attenuation for each. Remember that —6 dB equals half the reference voltage (the 2.0-MHz modulation should be used for reference).

Figure 13–22 represents the IF amplifier response curve of a color receiver. Thus, the curve frequencies are given in terms of typical IF range (41 to 47 MHz), and the VITS modulation frequencies are superimposed at corresponding points. For example, the 3.6-MHz VITS modulation occurs at the chroma or color frequency of 42.17 MHz.

In an ideal color receiver (and a good black and white receiver) the 2.0- and 1.5-MHz modulation signals should be at about the same level, the 0.5- and 3.0-MHz signals should be about 3 dB down from the 2.0-MHz reference

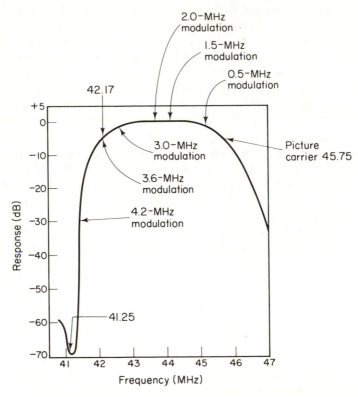

FIGURE 13-22 Typical color IF amplifier response.

level, the 3.6-MHz (chroma) signal should be about 6 dB down from (or about one-half) the 2.0-MHz reference level, and the 4.2-MHz modulation signals should be about 30 dB down from the 2.0-MHz reference.

13-6-4. Localizing Trouble with the VITS Waveform

To localize trouble, start by observing the VITS at the video detector. This localizes trouble to a point either before or after the detector. If the multiburst is normal at the video detector, check the VITS on other channels. If some channels are good but others are not, you probably have a defective tuner or antenna-system troubles. Do not overlook the chance of the antenna system causing "holes" (poor response on some channels). If the VITS is abnormal at the video detector on all channels, the trouble is in the IF amplifier or tuner. If the VITS is normal at the video detector for all channels, the trouble is in the video amplifier. Look for open peaking coils, off-value resistors, bad solder connections, partial shorts, and so on.

As an example of localizing trouble with the VITS, assume that you have a receiver with a poor picture (weak, poorly defined) but that the receiver is fully operative and that the antenna system is known to be good. Further assume that the VITS pattern at the video detector is about normal, except that the 2.0-MHz burst is low compared to the bursts on either side. This suggests that an IF trap is detuned, chopping out frequencies about 2 MHz below the picture carrier frequency.

Switch to another channel carrying VITS. If the same condition is seen, then your reasoning is good, and the IF amplifier requires adjustment. If the poor response at 2.0 MHz is not seen on other channels, possibly an FM trap in the tuner input is misadjusted, causing a drop in signal on only one channel. Other traps at the input (RF tuner) or the receiver could similarly be misadjusted or faulty.

One final note concerning VITS signals. The VITS pattern broadcast by all stations are not identical, although they all resemble those patterns shown here. This may cause a problem when switching from channel to channel during the trouble localization process. Be careful that a different VITS pattern does not lead you to believe a defect exists on one or more channels. Compare the VITS pattern with the picture. For example, if you get an odd VITS pattern on one channel and the picture is poor on that channel, you have definitely localized trouble to the tuner.

Index

Y

Y-axis, 6
Yoke tests, 325

Z

Z-axis, 8
Zener diode tests, 201, 207
Zero temperature coefficient (FET), 185